AIR, WATER AND SOIL POLLUTION SCIENCE AND TECHNOLOGY

AIR QUALITY OBSERVATION IN THE US

SYSTEMS, NEEDS, AND STANDARDS

AIR, WATER AND SOIL POLLUTION SCIENCE AND TECHNOLOGY

AIR, WATER AND SOIL POLLUTION SCIENCE AND TECHNOLOGY

AIR QUALITY OBSERVATION IN THE US

SYSTEMS, NEEDS, AND STANDARDS

MALCOLM PARISI
EDITOR

nova publishers
New York

For permission to use material from this book please contact us:
Telephone 631-231-7269; Fax 631-231-8175
Web Site: http://www.novapublishers.com

Additional color graphics may be available in the e-book version of this book.

Library of Congress Cataloging-in-Publication Data

ISBN: 978-1-63117-154-3

Published by Nova Science Publishers, Inc. † New York

CONTENTS

PREFACE

Accurate information about air pollutants in the troposphere, such as ozone, particulates, and their respective precursors, is essential to decision-making in domains as diverse as human health, environmental protection, climate change, and agriculture, as well as for assessments of air quality management strategies and policies. Atmospheric observations have only grown in importance as the risks posed by global climate change have come more clearly into focus, and observations will be critical in the years ahead both for tracking the impacts of mitigation strategies and predicting national needs in the realm of adaptation and preparedness. In this book the authors discuss air quality observation systems in the United States; ozone air quality standards; and the EPA's 2013 changes to the particulate matter standard.

Chapter 1 – Report of Committee on Environment, Natural Resources, And Sustainability of the National Science and Technology Council on Air Quality Observation Systems in the United States, dated November 2013.

Chapter 2 – The Environmental Protection Agency (EPA) is nearing the end of a statutorily required review of the National Ambient Air Quality Standards (NAAQS) for ozone. EPA projects proposal of a revised standard by December 2013, with promulgation late in 2014. NAAQS are standards for outdoor (ambient) air that are intended to protect public health and welfare from harmful concentrations of pollution. If the EPA Administrator changes the standard to a lower level, he or she would be concluding that protecting public health and welfare requires lower concentrations of ozone pollution than it previously judged to be safe. In high enough concentrations, ozone has been found to aggravate heart and lung diseases and may contribute to premature death. The ozone standards affect a large percentage of the population: as of December 2012, 123 million people (40% of the U.S.

population) lived in areas classified nonattainment for the primary (health-based) ozone NAAQS. If the standards are further strengthened, more areas would likely be affected, and those already considered nonattainment and sources that contribute to the nonattainment might have to impose more stringent emission controls. EPA last revised the ozone standard in March 2008, but the level chosen at that time remains subject to controversy. A 23-member panel of EPA science advisers, chosen from outside the agency, unanimously recommended a more stringent range of standards than the Administrator chose. With the change of Administrations in 2009, the agency agreed to reconsider the 2008 standard, but the process was short-circuited by a Presidential decision to await conclusion of the next regular review—the review now nearing completion. The agency begins a NAAQS review by compiling an Integrated Science Assessment, which summarizes the science surrounding the standards. The current assessment, released in February 2013, finds that the evidence associating ozone exposure with morbidity and mortality has strengthened since the 2008 review was completed. This would appear to support more stringent standards. Proposed standards might raise a number of issues, including whether their expected benefits justify their costs. This is a perennial issue raised by stakeholders when EPA considers revising the NAAQS, although, as the Clean Air Act is currently written, the agency is prohibited from weighing costs against benefits in setting these standards. The statute simply states that the Administrator is to set the primary standard at a level requisite to protect the public health, allowing an adequate margin of safety. A unanimous Supreme Court has found that the absence of language mentioning cost means that costs are not to be considered in setting these standards. Many in Congress would like to change this to require such a cost or cost-benefit consideration: in the 112th Congress, House-passed legislation would have done so. A related question is how nonattainment areas would lower emissions sufficiently to comply with a more stringent standard. Current federal standards for cars, trucks, nonroad vehicles and engines, power plants, and other stationary pollution sources are not strong enough to bring many areas into attainment, thus requiring local pollution control measures in those cases. To attain more stringent standards, there might be increased pressure to strengthen federal controls on the sources that contribute to ozone pollution; Congress might also revisit the perennial issue of how to control sources that contribute to pollution downwind, in other states. There are also monitoring issues: at present, only 675 of the nation's 3,000 counties have ozone monitors in place. EPA, the states, and Congress may consider whether the current monitoring network is adequate to detect violations of a more stringent standard; both EPA and state monitoring budgets are constrained, however, raising questions as to how any additional monitoring requirement would be funded. This report discusses the standard-setting process, the specifics of the current and most recent reviews, and issues that may be raised as EPA brings the current review to completion.

Chapter 3 – On January 15, 2013, the Environmental Protection Agency (EPA) published a final rule revising the National Ambient Air Quality Standard (NAAQS) for particulate matter (PM). The revised air quality standards were completed pursuant to the Clean Air Act (CAA) and, in part, in response to a court order and consent agreement. Based on its review of scientific studies available since the agency's previous review in 2006, EPA determined that evidence continued to show associations between particulates in ambient air and numerous significant health problems, including aggravated asthma, chronic bronchitis, non-fatal heart attacks, and premature death. Populations shown to be most at risk include children, older adults, and those with heart and lung disease, and those of lower socioeconomic status. EPA's review of and revisions to the PM NAAQS have generated considerable debate and oversight in Congress. The January 2013 revisions change the existing (2006) annual health-based ("primary") standard for "fine" particulate matter 2.5 micrometers or less in diameter (or $PM_{2.5}$), lowering the allowable average concentration of $PM_{2.5}$ in the air from the current level of 15 micrograms per cubic meter ($\mu g/m^3$) to a limit of 12 $\mu g/m^3$. The annual $PM_{2.5}$ NAAQS is set so as to address human health effects from chronic exposures to the pollutants. The existing "24-hour primary standard" for $PM_{2.5}$ that was reduced from 65 $\mu g/m^3$ to 35 $\mu g/m^3$ in 2006 was retained, as was the existing standard for larger, but still inhalable, "coarse" particles less than 10 micrometers in diameter, or PM_{10}. "Secondary" standards that provide protection against "welfare" (non-health) effects, such as ecological effects and material deterioration, are identical to the primary standards and the same as in 2006. The proposed rule published June 29, 2012, solicited comments on two options for a 24-hour $PM_{2.5}$ standard to improve visibility that were not adopted in the final rule. EPA revised the Regulatory Impact Analysis (RIA) accompanying its June 2012 proposed rule in part in response to comments received regarding the agency's cost and benefit estimates. In its December 2012 RIA, EPA estimated that the potential "quantifiable" health benefits (2010 $) associated with attaining the PM standard would range from $4.0 billion to $9.1 billion, and costs (2010 $) would range from $53.0 million to $353.0 million. Some stakeholders and some Members continue to express concerns that cost impacts would be more significant than those estimated by

EPA for those areas out of compliance with the new standards. EPA's revisions to the PM NAAQS do not directly regulate emissions from specific sources, or compel installation of any pollution control equipment or measures, but indirectly could affect operations at industrial facilities and other sources throughout the United States. Revising PM NAAQS starts a process that includes a determination of areas in each state that exceed the standard and must, therefore, reduce pollutant concentrations to achieve it. Following determinations of these "nonattainment" areas based on multiple years of monitoring data and other factors, state and local governments must develop (or revise) State Implementation Plans (SIPs) outlining measures to attain the standard. These often involve promulgation of new regulations by states, and the issuance of revised air permits. The process typically takes several years. Based on statutory scheduling requirements, nonattainment designations for revised PM NAAQS would not be determined until the end of 2014, and states would have until at least 2020 to achieve compliance with the January 2013 revised $PM_{2.5}$ NAAQS.

In: Air Quality Observation in the US
Editor: Malcolm Parisi
ISBN: 978-1-63117-154-3

Chapter 1

AIR QUALITY OBSERVATION SYSTEMS IN THE UNITED STATES*

Richard Scheffe and Joseph Tikvart

NATIONAL SCIENCE AND TECHNOLOGY COUNCIL WASHINGTON, D.C. 20502

Dear Colleagues:

We are pleased to transmit the report "Air Quality Observation Systems in the United States." This report, prepared by the Air Quality Research Subcommittee of the National Science and Technology Council's Committee on Environment, Natural Resources, and Sustainability, describe the Nation's interlinking networks of ambient air observation programs; explores opportunities to maximize the value of these programs; identifies common data needs and uses; and considers approaches to leveraging these important resources across multiple Federal agencies.

This report complements findings and recommendations in a number of National Research Council reports, including *Earth Science and Applications from Space: National Imperatives for the Next Decade and Beyond* (2007) and *Observing Weather Climate from the Ground Up: A Nationwide Network of*

* This is an edited, reformatted and augmented version of the Committee on Environment, Natural Resources, And Sustainability of the National Science and Technology Council, November 2013.

Networks (2009); includes foundational information for integration into the National Plan for Civil Earth Observations, currently in development: and is anticipated to serve as a valuable resource as the Nation implements President Obama's recently released Climate Action Plan.

Sincerely,

Tamara Dickinson, OSTP (Co-Chair, CENRS) Date 11/22/13
Glenn Paulson, EPA (Co-Chair, CENRS) Date 11/12/13
Kathryn Sullivan, NOAA (Co-Chair, CENRS) Date 11/20/13

EXECUTIVE SUMMARY

In June 2013, President Obama released his Climate Action Plan. The Plan calls for a range of activities to help mitigate the impacts of carbon pollution, prepare for and adapt to those impacts that cannot be avoided, and enhance international efforts to reduce the harmful global effects of climate change. Success in each of the Plan's three components will require access to and application of the highest quality atmospheric science.

Fortunately, as a result of decades of attention to the issue of air quality generally – from ozone and carbon dioxide to soot and other particulates – the United States is in good position to address the added observational and modeling challenges posed by climate change. Federal agencies and state and local partners invest hundreds of millions of dollars annually to maintain and operate the Nation's ground- and space-based air quality observation networks, and to conduct short-term field studies relating to air quality. The data collected by these activities have been indispensable to the effective development and implementation of policies to protect public health and the environment, and will be even more so as the Nation addresses the emerging impacts of climate change.

This report catalogs the wide range of U.S. air quality measurement modalities and programs, including routine regulatory and deposition networks, intensive field studies, satellites, and fixed-site special-purpose networks operated or overseen by Federal departments and agencies – including the Environmental Protection Agency (EPA), National Oceanic and Atmospheric Administration (NOAA), National Aeronautics and Space Administration (NASA), U.S. Department of Agriculture (USDA), Department of Energy (DOE), Department of Health and Human Services

(DHHS), Department of Homeland Security (DHS), and Department of the Interior (DOI) – and by various state, local and tribal partners. It also highlights some leading observational needs and opportunities, and identifies some of the barriers to fulfilling those needs and leveraging those opportunities. Specific programs and the pollutants they measure are documented in detail in the appendices. Among this report's findings (not ranked in priority order) are:

Observational Needs

1. Although local and regional observation and emissions reduction efforts are generally well advanced in the United States, there is a need to improve characterization of pollutants from distant/international sources to better evaluate their contribution to domestic air quality.
2. There is a need for enhanced atmospheric, deposition, and effects monitoring to understand deposition impacts on aquatic and terrestrial ecosystems.
3. There is a growing need to document global changes in climate that have the potential to affect the natural release or atmospheric processing of pollutants.
4. There is a need for greater attention to precursors of ozone and particulate matter in order to accurately assess the success of emissions control strategies.
5. There is an ongoing need for more cost effective means of compliance monitoring, to reduce financial burdens on state and local governments.
6. There is a need for vertically resolved observations of ozone, particulate matter (including its composition), and their precursors, to evaluate and improve air quality modeling.

Opportunities

1. Satellite sensing of air quality and emissions is rapidly maturing in its capability to augment and extend the spatial and temporal coverage offered by fixed-site monitoring networks and short-term special studies, allowing for unprecedented, high value ventures into satellite-

based, ecosystem-characteristic measurements. U.S. agency geostationary missions in the near future will provide coverage at scales highly relevant to urban air quality, presenting notable future opportunities for analysis and integration.
2. Air quality models are increasingly able to reliably augment direct observations with credible spatial, temporal, and compositional information lacking in direct measurements.
3. Open access policies are providing enhanced access to observational data, metadata, and processing tools, facilitating harmonization and consolidation of disparate datasets and collaboration among previously unrelated research teams.

Barriers to Progress

1. Funding for long-term maintenance and updates of observation infrastructure is often subject to annual budgeting processes that inhibit long-range planning.
2. Agencies often support their own priority programs rather than pooling resources on joint efforts in domains of shared responsibility.
3. Some issues are not adequately addressed because they do not fall clearly within any one entity's regulatory purview.
4. Funding is often focused on the development of advanced technologies without adequate accompanying support for the development of tools for processing and transferring the resulting data streams.
5. In the absence of government mandates or other incentives, commercialization of air quality monitoring technologies suffers from weak market incentives.

Conclusion

The United States has a robust and invaluable network of air quality observation systems, and recent improvements in technology are providing unprecedented opportunities to enhance current capabilities. However, current funding patterns and limited coordination among departments and agencies with equities in air quality observations are preventing the Nation from reaching its full potential in this important domain.

The NSTC Committee on Environment, Natural Resources, and Sustainability (CENRS) should consider establishing a standing, multi-agency Observations Working Group under the Air Quality Research Subcommittee (AQRS) whose responsibilities could include:

a. conducting periodic adequacy reviews of the Nation's air quality observational capabilities;
b. identifying and addressing gaps and overlaps among programs;
c. building cooperation and coordination among government programs;
d. advising on minimum standards for program design/implementation;
e. promoting the use of common data formats and communications protocols;
f. identifying opportunities for development of, and review/ recommending the use of, new observational technology;
g. fostering data quality elements, monitoring across appropriate gradients, and the role of modeling;
h. identifying low-priority networks that can be discontinued and defunded in order to offset the costs of new investments.

The Working Group could also develop strategies to address such needs as:

a. initiating monitoring of reactive gas and particulate nitrogen compounds, which are precursors of ozone and particulate matter, contributors to acid deposition, and nutrients in ecosystems;
b. collocating instrumentation at core measurement sites to facilitate inter- comparison with satellite observations;
c. targeting observations in rural/remote areas to measure regional backgrounds and contributions from long-range transport of pollutants;
d. establishing more robust air toxics monitoring near major industrial facilities to help investigate whether air toxics emissions are associated with reported human health effects in nearby communities; and
e. targeting intensive field studies designed to elucidate critical processes that determine atmospheric concentrations of ozone and particulate matter and other air pollutants.

INTRODUCTION

Accurate information about air pollutants in the troposphere, such as ozone, particulates, and their respective precursors, is essential to decision-making in domains as diverse as human health, environmental protection, climate change, and agriculture, as well as for assessments of air quality management strategies and policies. Atmospheric observations have only grown in importance as the risks posed by global climate change have come more clearly into focus, and observations will be critical in the years ahead both for tracking the impacts of mitigation strategies and predicting national needs in the realm of adaptation and preparedness.

Modeling of the emissions and processes that affect air quality is improving rapidly. But direct observation remains the gold standard for tracking man-made and naturally occurring atmospheric species, the spatial and temporal distribution of which are governed by such complex influences as: the sources of those species; relevant meteorology; physical and chemical processes that transport, transform, and remove various species; and exchanges between the atmosphere and terrestrial and aquatic environs.

Reliable assessments of atmospheric composition are especially important given the high financial stakes; the costs and benefits of Clean Air Act requirements alone are estimated to be on the order of tens of billions of dollars per year (National Research Council, 2004). In response to such regulatory and other incentives, a wide variety of Federal agencies and organizations make air quality observations or process or take advantage of air quality data, including EPA, NOAA, NASA, USDA, DOE, DHHS, DHS, and DOI, as do associated state, local and tribal partners. In many cases, these agencies share common information needs, and the value of air quality observations could be enhanced by coordinated planning or shared operations. Opportunities also exist to increase the value of air quality observations by integrating them across environmental media, pollutant categories, and spatial scales.

This report (1) describes the measurement parameters, locations, and sponsoring organizations of ambient monitoring and other air quality observation and assessment efforts, focusing primarily on U.S. assets; and (2) identifies observational gaps and opportunities to enhance the value of existing measurement programs through inter- agency cooperation and collaboration. Appendices provide a comprehensive inventory of monitoring networks, observation programs, and pollutants measured.

Current and Emerging Air Quality Assessment Challenges

Over the last two decades, air quality management in the United States has focused on regional-scale air pollutants such as ozone, particulate matter, and acid deposition, all of which remain issues of concern and are likely to continue as such for the foreseeable future, especially if air quality standards continue to tighten.

Traditional management strategies typically involve independent, specific approaches based on targeted monitoring of individual pollutants. Increasingly, however, air quality management requires a more comprehensive and well-integrated assessment framework:

- Multiple pollutants – Typically, multiple pollutants are emitted by individual emission sources and participate in interrelated atmospheric, chemical, and physical transformation and loss processes. Too often, however, measurement systems are still structured to support only single-pollutant assessments.
- Multiple environmental media – The atmosphere is closely coupled with terrestrial and aquatic systems. These systems are major sinks for air pollutants, leading to effects on ecosystems from acids, nutrients, and toxics. In turn, soils, vegetation, and aquatic systems re-emit mercury and persistent organic pollutants (POPs), and meteorology and climate affect biogenic and biomass burning emissions. These linkages require a broader perspective on environmental monitoring, which traditionally addresses issues on an isolated, single-media basis. Coordinated monitoring across multiple media is required to accurately assess progress mitigating the effects of atmospheric pollution on human health, ecosystems, and agriculture.
- Multiple spatial scales – Long-range (inter-regional and intercontinental) pollutant transport is becoming increasingly important as transport across U.S. borders increases from expanded world development and as local emissions decline (NRC, 2009a). Meanwhile, urban-based field studies have demonstrated high pollutant levels in the near-source/roadway environment, where a majority of the North American population lives and the chemical environment is dynamic and poorly understood. These scale issues, at opposite ends of the spatial spectrum, challenge the current assessment framework that emphasizes regional air quality management.

- Climate-air quality interactions – The bi-directional interaction between air quality and climate change impacts air quality management. A variety of emissions, atmospheric chemistry, and transport processes that affect air quality are modified by climate change. Conversely, several air pollutants, particularly greenhouse gases, ozone, and black carbon particulate matter, are significant climate forcers, and air quality changes impact atmospheric and emissions processes, impacting climate. Moreover, climate forcers and conventional air pollutants are largely emitted from common sources. Consequently, emerging energy policies designed for moderating climate and policies designed to improve air quality are intrinsically connected, and measurement-system design should reflect this relationship.

Addressing these challenges would also address several challenges highlighted in integrated observational strategies (see Appendix A) and in three reports by the National Academies.

In *Global Sources of Local Pollution: An Assessment of Long-Range Transport of Key Air Pollutants to and from the United States* (NRC, 2009a), improved satellite observations, in situ monitoring, and intensive field campaigns were all highlighted as ways to improve understanding of air pollution transport. In *Air Quality Management in the United States* (NRC, 2004), the highest priority recommendation for improving air quality management was to "strengthen the scientific and technical capacity of the [air quality management] system to assess risk and track progress." Carefully designed and maintained monitoring can contribute to this progress by improving understanding of ambient concentrations, emissions, transport, and deposition and by improving modeling through more complete data for model evaluation.

A third National Academies report, *Observing Weather and Climate from the Ground Up: A Nationwide Network of Networks* (NRC, 2009b), cited four types of observations as the "highest priority observations needed to address current inadequacies," all of which are relevant for air quality monitoring:

- Height of the planetary boundary layer
- Soil moisture and temperature profiles
- High-resolution vertical profiles of humidity

- Measurements of air quality and related chemical composition above the surface layer.

Uses of Air Quality Observations

The air quality observation systems discussed in this report serve four broad disciplines: (1) human exposure and health effects, (2) ecosystem exposure and effects, (3) air quality management efforts, and (4) basic atmospheric science, including linkages between air quality and climate change. Although disciplines (1) and (2) are described in general terms below, this report does not attempt to elucidate the major needs or gaps of U.S. observation programs focused specifically on human exposure and health effects or ecosystem impacts—including studies of physical or chemical properties of aquatic and terrestrial systems—which generally need especially high-resolution data over multiple observational axes.

Human Exposure and Health Effects

Relating health effects to observed concentrations of pollutants is a critical use of air quality observations. Forecasting of future air quality conditions allows the general public, particularly sensitive groups, to modify behavior and protect their health by reducing potential exposure to poor air quality. Verification and improvement of model predictions and interpretative forecasts relies on timely availability of accurate and precise observations. In addition to surface observations of criteria pollutants, such as ozone and particulate matter, measurements of their precursors, particulate matter composition, and the vertical distribution of pollutants are needed to improve model predictions.

Many epidemiological studies of the health effects of air pollution use data from existing monitoring networks. For example, Pope et al. (2009) used measurements from EPA monitoring networks to show that decreases in $PM_{2.5}$ concentrations lead to increased life expectancy. In general, however, the use of existing monitoring networks for epidemiological studies of air pollution poses challenges. Monitoring networks typically involve air quality observations from relatively small numbers of instruments in metropolitan areas, placed in areas that may not capture the full range of concentrations to which people within the area are exposed. Likewise, monitoring networks of criteria pollutants are designed to provide pollutant levels over an averaging time defined by the National Ambient Air Quality Standards (NAAQS), which can prove limiting for some health studies.

For measurements not directly related to compliance with air quality standards, such as air toxics or speciation of particulate matter, the frequency of measurements can be considerably lower than ideal for epidemiological studies.

Further, when data from an existing monitoring network are used to study health effects, the research is limited to the chemical species being measured by the network, as well as by pre-determined measurement periods and spatial coverage.

It is important to note that health impacts are the primary inputs into the evolution of NAAQS, which in turn influence monitoring design. Other types of health impacts research (e.g., for air toxics) complement the epidemiological studies that use ambient monitors, but the inter-relationship of the monitoring design and health effects studies illustrates the need for careful design of ambient monitoring programs. This network design should generally consider the same scale (e.g., local, regional) as the epidemiology studies used to set the health-based standards.

Ecosystem Exposure and Effects

Atmospheric observations are a major part of ecological effects assessment efforts, either directly or through air quality models. Watershed acidification, eutrophication, and direct damage to vegetation are examples of major ecosystem welfare issues whose understanding can be enhanced by atmospheric characterization studies. The ecosystem assessment community is an important client of air quality observations, using available data as inputs to ecosystem exposure models and as trend indicators relating the effectiveness of emissions strategies on atmospheric deposition. The characterization demands for ecosystem analyses may be as demanding as those associated with human health and exposure communities, given the spatial heterogeneity of vegetation, soil types, and microclimates within and across watersheds and ecosystems that affect, and are affected by, atmospheric deposition. Furthermore, most U.S. monitoring stations are distributed according to population-weighted criteria, creating major information gaps in sensitive ecosystems.

Air Quality Management Efforts

Air quality management practice includes the establishment of human and ecosystem health-based standards and the subsequent development of rules, programs, and implementation steps designed to achieve the emission changes needed to meet air quality targets. The cyclic nature of air quality management reflects both the evolution of air quality standards (based on improving knowledge of the effects of air pollution) and evaluation of whether implemented programs produced intended results. In recent years, air quality management has sought more direct evidence of the connections along the source-to-effects continuum, and of the relationship between emissions changes and air quality improvements, to better assess the effectiveness of emissions strategies.

The regulatory nature of air quality management decisions places special demands on the observations used to support this work. Ambient monitoring of criteria air pollutants such as ozone, particulate matter, etc., is designed to be used to determine if an area is compliant with a specified NAAQS. Only certified measurements from Federal Reference or Equivalent Methods (FRM/FEM) can be used for comparison to the NAAQS. However, FRM/FEMs have not been established for toxic air pollutants, in part because the regulatory structure of the Clean Air Act (CAA) does not require ambient monitoring for these pollutants. In practice, the reliance on FRM/FEM certification and a lack of regulatory drivers has hindered development and commercialization of potentially useful air quality monitoring technologies.

Basic Atmospheric Science, Including Linkages between Air Quality and Climate Change

Air quality data can improve the characterization of physical and chemical processes underlying pollutant release, transformation, and removal and can inform the development, evaluation, and refinement of quantitative atmospheric chemistry models. Such models, in turn, can be critical to air quality management efforts and to assessments of air quality on health, ecosystems, and climate. And while models are essential for making predictions, such as air quality forecasts, climate predictions, and estimated impacts of future emissions changes, they are also key to characterizing the multi-dimensional (space, time, and species) chemical state of the current atmosphere and to understanding historical conditions.

OVERVIEW OF OBSERVATION PROGRAMS

A variety of measurement programs support air quality assessments. These include:

- routine regulatory and deposition networks
- intensive aircraft and ground-based field studies
- radiosonde programs
- satellite measurements
- ground-based remote-sensing networks
- focused, fixed-site, special purpose networks

The following inventory of representative programs is not intended to be comprehensive but focuses on extant, routinely operating North American networks, with some mention of European and international efforts relevant to North American assessments. More detailed information is available in the appendices.

Routine Surface-Based Ambient Air and Deposition Networks

Routine ambient air and deposition monitoring networks in North America comprise more than 3000 fixed platforms (Figure 1), measuring numerous gaseous species and aerosol properties (Appendices B and C). Many of these longstanding U.S. networks were required or catalyzed by the 1970 Clean Air Act (CAA), subsequent CAA amendments, National Ambient Air Quality Standard (NAAQS) reviews, and/or National Academy of Sciences (NAS) recommendations. Examples include the Clean Air Status and Trends Network (CASTNET) and National Atmospheric Deposition Program (NADP) addressing acidification; the Photochemical Assessment Monitoring Stations (PAMS) in response to persistent ozone pollution; and the $PM_{2.5}$ monitoring networks following promulgation of the 1997 NAAQS. Federal regulations describe how sites in these networks are to be located, and the Federal Reference Methods (FRM) or Federal Equivalent Methods (FEM) for the measurements made at many of the required sites.

Most routine air quality monitoring stations in the United States are owned and operated by nearly 300 state and local government and Tribal agencies. These state and local air monitoring stations (SLAMS) are the principal source of ambient measurements of the six criteria air pollutants (ozone, nitrogen dioxide, carbon monoxide, sulfur dioxide, lead, particulate matter - PM_{10} and $PM_{2.5}$), each of which has one or more NAAQS specifying a concentration level and averaging period (http://www.epa.gov/ttn/naaqs/). Most of these networks also include stations operated by Federal agencies, typically in rural / remote sites. These networks are indirectly supported by extensive meteorological networks (Appendix D).

National air monitoring regulations for U.S. programs are codified in Parts 50, 53, and 58 of Title 40 of the Code of Federal Regulations (CFR).

Funding for these programs is through CAA Sections 103 and 105 concerning Federal grants to agencies and tribes. States and local agencies are required to match Federal Section 105 contributions.

Criteria Gas and Ozone Precursor Monitoring

Criteria Gas Networks

Approximately 1500 surface stations measure some combination of criteria gases, with nearly 1100 of these stations measuring ozone, using FRMs or FEMs. Several hundred monitors report concentrations for CO, SO_2, and NO/NOx. The majority of these stations are SLAMS, although Federal agency networks such as CASTNET, National Park Service (NPS) monitors, and a variety of special purpose monitors provide additional coverage (Appendices B and C). CASTNET and NPS provide the majority of rural criteria pollutant platforms.

Photochemical Assessment Monitoring Stations (PAMS)

Approximately 75 sites in 22 cities were deployed by state and local agencies in the early 1990s to measure ozone precursors, largely in response to a 1991 National Academy of Sciences study (NRC, 1991). PAMS and the air toxics network (see below) provide the majority of routinely available non-methane organic carbon (NMOC) measurements. A number of ozone precursor C2-C10 alkanes and alkenes, aromatics, formaldehyde, and acetaldehyde are measured using a combination of continuous methods and sampling techniques over 3- and 24-hour collection periods, often limited to the ozone season (April – October). The 1990 CAA Amendments required areas classified as serious and above with respect to the contemporary (1990-1992) ozone NAAQS to implement PAMS, which has undergone minor modifications since then. Most volatile organic compound (VOC) sampling sites include instrumentation for O_3 and NO/NOx.

Particulate Monitoring

$PM_{2.5}$, PM_{10}, and $PM_{10\text{-}2.5}$ Mass Networks

The 1997 promulgation of a fine particulate NAAQS (EPA, 1997) led to deployment of over 1500 $PM_{2.5}$ sites (about 1000 currently), used to determine whether an area complies with the standard. These sites use an FRM or FEM, sampling over 24 hours daily, or every third or sixth day. Nearly 300 additional measurements not meeting FRM or FEM specifications are provided by the chemical speciation sites (see below). Approximately 600 stations provide indirect measurements of continuous (hourly resolution) $PM_{2.5}$ mass, using a variety of techniques. Continuous $PM_{2.5}$ mass measurements have been granted FEM status based on revised monitoring regulations issued in 2006 (EPA, 2006) that provided new approaches for demonstrating equivalency.

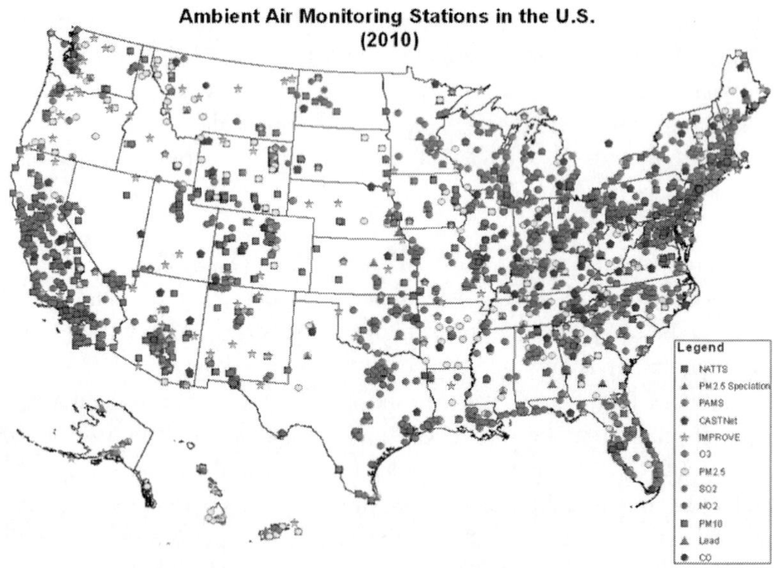

Figure 1. Aggregate map of the majority of routine U.S. monitoring stations, illustrating relatively broad coverage across the continental United States. Note spatial gaps in sparsely populated areas.

Approximately 1000 PM_{10} samplers (24-hr sampling period, typically collected every sixth day) remain in operation. Although a $PM_{10\text{-}2.5}$ standard has not been promulgated, EPA developed a $PM_{10\text{-}2.5}$ FRM based on mass difference of concurrent PM_{10} and $PM_{2.5}$ measurements. $PM_{10\text{-}2.5}$ measurements are incorporated in the NCore network (see below).

Interagency Monitoring of Protected Visual Environments (IMPROVE) Program

The IMPROVE network, with over 100 sites, has provided nearly a two-decade record of major components of $PM_{2.5}$ (sulfate, nitrate, organic and elemental carbon fractions, and trace metals) in pristine areas of the United States (Figure 2). IMPROVE is led by the National Park Service, with various other Federal and state agencies providing support operations. The primary focus of the network is to track visibility and trends in visibility.

$PM_{2.5}$ Chemical Speciation Monitoring

In addition to the IMPROVE network, over 300 EPA speciation sites were added from the years 2000 - 2002 in urban areas of the United States to assist $PM_{2.5}$ assessment efforts. No FRM exists for particulate speciation, which is not directly required to determine attainment, and there are slight differences among the different monitors used in the Chemical Speciation Network (CSN).

However, the network's coverage (Figure 2) across urban and rural areas has proven essential for a wide range of research and analysis. The speciation networks typically collect a 24-hour sample every three, and sometimes six, days. Daily 24-hour speciation collection is limited to occasional efforts in the SEARCH (see below) network. Similarly, only a handful of sites provide near continuous speciation data, usually limited to some combination of sulfate, carbon (organic and elemental splits), and nitrate. This enables insights into diurnal speciation patterns, helpful in diagnosing various cause-effect phenomena related to emissions characterization, source attribution analysis, and model evaluation. In addition, the National Air Toxics Trends Stations (NATTS; see below) include near continuous aethalometer measurements based on optical absorption as a surrogate for absorbing aerosol (e.g., elemental carbon).

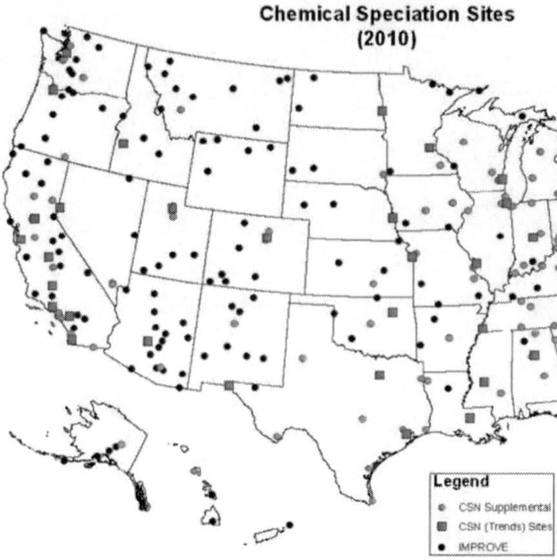

Figure 2. Locations of chemical speciation sites delineated by program type.

SouthEastern Aerosol Research and Characterization (SEARCH) Study

SEARCH is an industry-funded network that originally emerged from the Southern Oxidants Study (SOS) in the 1990s. It has operated for over a decade in response to 1997 revisions to the NAAQS for ground-level ozone and particulate matter. SEARCH provides an array of standard criteria pollutant measurements, in addition to daily PM speciation at selected times and locations, gaseous ammonia (NH_3), total peroxyacetyl nitrate (PAN), nitric acid (HNO_3), reactive oxidized nitrogen (NOy), and true nitrogen dioxide (i.e., a measurement of NO_2 concentration unaffected by other nitrogen oxides, which contaminate FRM NO_2 measurements).

These measurements typically had not been available in major government-funded routine networks. Measurements of pollutant composition, as well as mass, are necessary in order to identify sources of ozone precursors and fine particulate matter, and to attribute health effects to specific components.

PM Supersites Program

This program (Solomon et al., 2008) provided highly resolved aerosol measurements at eight U.S. cities for several time periods from 1999 through 2004, with some sites collecting data after 2004. A number of instrument configurations were deployed, ranging from additional locations for standard speciation monitors, to systems capturing near-continuous size-dependent speciation profiles.

The National Core (NCore) Network

NCore (Scheffe et al., 2009) is an 80-site multiple pollutant supplement to the routine monitoring networks. It was fostered by the Ambient Air Monitoring Strategy for State, Local, and Tribal Air Agencies (EPA, 2008), and promulgated in the 2006 Code of Federal Regulations as part of the new monitoring rule (EPA, 2006). NCore is designed to capture urban- and regional-scale representative concentrations of a variety of trace gases (CO, SO_2, NOy, and NO) and aerosols (PM_{10} and $PM_{2.5}$ mass and chemical speciation) to support a range of health effect, model evaluation, and other research studies. The NCore sites are designated multiple pollutant sites that require co-location with existing $PM_{2.5}$ chemical speciation measurements. Deployment of the network took place through 2011.

Air Toxics Monitoring Program

National Air Toxics Trends Stations (NATTS) Network

State and local agencies have measured a variety of metallic and gaseous toxic air pollutants at over 200 locations since the 1980s. Broad access to and use of these data was hampered by a lack of centralized databases and multiple sampling and laboratory protocols, creating data quality and consistency concerns. To address these inconsistencies, the NATTS network was conceived in 2001, consisting of 27 sites. The sampling protocol typically has been every sixth day for 24 hours.

Among the priority ranked 33 air toxics of U.S. concern, observations of benzene and other common aromatics are generally widespread and relatively reliable. However, other potentially important air toxics are less well represented in air monitoring. During the initial start-up of the NATTS, six priority pollutants (formaldehyde, benzene, 1,3- butadiene, hexavalent chromium, acrolein, and arsenic) were targeted for inclusion, based on results of the 1996 National Air Toxics Assessment (NATA; http://www.epa.gov/ttn/atw/natamain/index.html).

Building on efficiencies in methodologies and the 1999 and 2002 NATAs, NATTS observations expanded to include the following:

- Gas-phase compounds: acetaldehyde, acrolein, benzene, carbon tetrachloride, chloroform, dichloropropane, dichloromethane, formaldehyde, naphthalene, perchloroethylene, trichloroethylene, vinyl chloride, and 1,3-butadiene.
- Metals in PM_{10}: nickel, arsenic, cadmium, manganese, beryllium, and lead.
- Total suspended particle (TSP) mass: hexavalent chromium.
- Combined gas-phase and TSP: naphthalene and benzo(a)pyrene.
- Light absorbing carbon through aethalometers at a subset of sites.

In 2009, the EPA initiated a school air toxics study (http://www.epa.gov/schoolair/) to sample and analyze for a variety of hazardous air pollutants (HAPs; "air toxics") at 63 schools in 22 states. The Integrated Atmospheric Deposition Network (IADN) program (see below) also analyzes for selected metals (As, Pb, Cd, and Se) and other air toxics, while the National Atmospheric Deposition Program (NADP) includes mercury (Hg) monitoring in some sites. These programs are discussed below. The spatial and temporal patterns of many HAPs have strong localized signals associated with near field source emissions.

This makes characterization of air toxics concentrations especially challenging, given both the number of chemical species of concern and the difficulty in accounting for impacts of a myriad of emission sources. Consequently, the U.S. EPA has primarily relied on air quality modeling to communicate risks associated with air toxics in the NATA reports.

Deposition Networks

Precipitation-Based Networks: NADP and IADN

Precipitation chemistry is an important link between the atmosphere and terrestrial and aquatic systems. The National Atmospheric Deposition Program (NADP) currently consists of five subnetworks. Two of these subnetworks, the National Trends Network (NTN) and the Atmospheric Integrated Research Monitoring Network (AIRMoN), oversee approximately 250 sites that analyze for ions that have significant acidification and eutrophication effects. AIRMoN provides precipitation samples that facilitate temporal resolution. NTN data have described trends in precipitation chemistry across the United States since about 1978.

Two additional NADP subnetworks measure mercury. The Mercury Deposition Network (MDN) includes over 90 sites that measure total mercury in precipitation. The Atmospheric Mercury Network (AMNet) provides data on the atmospheric concentrations of mercury in gaseous and particulate forms, and other data needed to estimate dry deposition at twenty sites across North America. Monitoring mercury in the atmosphere is important for model evaluation and tracking the atmospheric response to emissions reductions.

AMNet, which began as a pilot partnership networking and standardizing previously deployed instruments, was formally adopted by NADP in 2009. The joint Canadian- United States Integrated Atmospheric Deposition Network (IADN) includes a mix of stations across the Great Lakes that sample both precipitation and ambient air for a range of toxic compounds. IADN emphasizes measurement of many of the more persistent organic compounds, including polychlorinated biphenyls (PCBs), pesticides, dioxins, and toxic metals (lead, cadmium, arsenic, and selenium).

Clean Air Status and Trends Network (CASTNET)

CASTNET was established in the early 1990s to track changes in dry deposition of major inorganic ions (nitrate and sulfate) and gaseous precursors associated with CAA Title 4 reductions in sulfur and nitrogen, designed to address surface water acidification in eastern North America. The network of over 80 sites has expanded from an eastern United States focus to cover large areas in the West. CASTNET provides weekly averaged ambient measurements of major ions (sulfate, nitrate, and ammonium) and gaseous sulfur dioxide and nitric acid. A subset of sites includes ozone and IMPROVE $PM_{2.5}$ speciation instruments. CASTNET site locations were designed to reflect regional scale air mass samples, relatively free from local urban source signals. The ambient concentrations are used in algorithms that estimate deposition velocity to calculate dry deposition.

The Ammonia Monitoring Network (AMoN) is the only network providing a consistent, long-term record of ammonia gas concentrations across the United States. AMoN is an approved NADP network and uses passive samplers collecting weekly integrated measurements, located at nearly 50 of the CASTNET locations.

Other Air Monitoring Networks

For completeness, European air monitoring networks and national/ international networks for monitoring persistent organic pollutants (POPs) are listed respectively in Appendices E and F.

Accessing Surface Network Data

Access to routine measurements is available through the following portals:

- EPA's Air Quality System (http://www.epa.gov/ttn/airs/airsaqs/) and related DataMart (http://www.epa.gov/ttn/airs/aqsdatamart/) house criteria gas, PAMS, PM mass, PM speciation, and air toxics data.
- EPA's AIRNow (http://www.airnow.gov/) and AIRNowTech (http://www.airnowtech.org/) provide near real time access to ozone and continuous $PM_{2.5}$ mass data.
- VIEWS (Visualization Information Exchange Web System - http://vista.cira.colostate.edu/views/ developed by the Regional Planning Organizations (RPOs) in support of visibility assessments) houses IMPROVE and EPA $PM_{2.5}$ speciation data.
- CASTNET (http://www.epa.gov/castnet/), NADP (http://nadp.sws.uiuc.edu/), and IADN (http://www.epa.gov/glnpo/monitoring/air2/index.html) provide direct access to deposition data. NADP data includes sub-networks (NTN, MDN, AMNet, AMoN) and AIRMoN (http://nadp.sws.uiuc.edu/airmon/).
- The Health Effects Institute (HEI) air quality database provides access to, and analysis tools for, processed $PM_{2.5}$ chemical speciation data (http://www.healtheffects.org/research.htm).
- Supersites Integrated Relational Database (SIRD) is described at http://www.epa.gov/ttn/amtic/ssdatamg.html.
- SouthEastern Aerosol Research and CHaracterization (SEARCH) is described and data availability identified at http://www.atmospheric-research.com/studies/SEARCH/index.html.
- Interagency Real-Time Smoke Particulate Monitoring provides real-time smoke concentration data (along with other meteorological information) from portable smoke monitors, and is described at http://app.airsis.com/USFS/.

Intensive Field Campaigns

Intensive field campaigns (Appendix G) of relatively short duration supplement routine long term monitoring networks by measuring spatial, temporal, and compositional distribution of pollutants and precursors.

These studies are designed to investigate the emission and physical and chemical processing of precursors and pollutants to understand their source, fate, transport, and removal. Typically, field campaigns use some combination of aircraft- and/or ship-based studies, satellite- and ground-based remote sensing, research-grade instrumentation, and advanced analytical methods. These efforts complement routine ground-based measurements, which usually do not address reactive gaseous species, aerosol size distributions, organic chemistry characterization, and vertically stratified data.

There is a long history of intensive field campaigns, starting with the Regional Air Pollution Study (RAPS) in the 1970s which formed the basis for evaluating the early photochemical models used in ozone assessments. Landmark campaigns in the United States through the 1980s and 1990s were reviewed as part of the 2000 NARSTO ozone assessment (Solomon et al., 2000), including the Southern California Air Quality Study, the San Joaquin Valley Air Quality Study (SJAQS)/Atmospheric Utility Signatures, Predictions, and EXperiments (AUSPEX), and the Southern Oxidant Study (SOS). Over the last decade, there has been a series of field campaigns focusing on characterization of surface-level aerosols through the PM Supersites program (Solomon et al., 2008).

While early campaigns focused on urban environments, the Eulerian Model Evaluation Field Study (EMEFS) and SOS during the early 1990s shifted the focus toward regional spatial scales. This was consistent with the dominant air pollution concerns of the time, notably acid rain and ozone.

In addition to addressing urban areas of concern, such as Houston and Los Angeles, more recent campaigns have extended spatial scales beyond regional studies to address long-range transport and continental-scale atmospheric processes. Some of these campaigns include local and regional studies for the northeast and southeast United States, portions of Texas, and central and southern California, and intercontinental studies of transport across North America and the Atlantic, Pacific, and Indian Oceans. A variety of Federal agencies (particularly NOAA and NASA) and state entities have served as leads for these studies. Appendix G provides a listing of key studies conducted since the late 1990s, with important earlier campaigns identified in footnotes. Several recent, highly relevant, campaigns are briefly described below.

The Intercontinental Transport and Chemical Transformations project of 2002 (ITCT-2k2) investigated springtime transport along the Pacific coast of North America. The campaign combined ground- and aircraft-based measurements with model simulations and satellite data products. The focus was on tropospheric chemistry and transport of ozone, fine particles, and chemically-active greenhouse compounds. The study shed light on the intercontinental transport of ozone and aerosols, and the impacts this transport has on local air quality and climate.

In 2004, the International Consortium for Atmospheric Research on Transport and Transformation (ICARTT) served as an organizing umbrella for North American and European field campaigns addressing regional scale processes in both continents, as well as trans-Atlantic transport phenomena (Fehsenfeld et al., 2006). The North American studies included the Intercontinental Chemical Transport Experiment - North America (INTEX-NA 2004) and the New England Air Quality Study - Intercontinental Transport and Chemical Transformation (NEAQS – ITCT 2004) programs. Analysis was based on a variety of satellite, aircraft, ship-based, and ground-based measurements. The ICARTT campaigns provided insights into trans-Atlantic processing of ozone precursors, lightning-generated NO_x emissions, secondary organic aerosol processes, and biomass burning.

The ICARTT campaigns had been preceded by the North American Regional Experiment (NARE) in the 1990s, which studied synoptic scale transport in the North Atlantic (Fehsenfeld et al., 1996; Penkett et al., 1998). The Transport and Chemical Evolution over the Pacific (TRACE-P) campaign of 2001 catalyzed much of our current understanding of Asian outflow to North America. The INTEX-NA mission was followed by the 2006 Intercontinental Chemical Transport Experiment (Phase B) (INTEX-B) aircraft mission, which studied pollutant transport flow across the north Pacific and into the western United States. INTEX-B was also linked with the 2006 Megacity Initiative: Local and Global Research Observations (MILAGRO) mission, which studied pollutant outflow from Mexico City. Most of the large intercontinental scale field campaigns are considered key parts of the IGAC program (Appendix A). Findings specific to Northern Hemisphere transport have been synthesized by the Hemispheric Transport of Air Pollution (HTAP) task force (Keating and Zuber, 2007).

The Texas Air Quality Study (TexAQS and TexAQS II) during 2000 and 2006 involved intensive research campaigns designed to address some of the unique VOC chemistry and transport features of southeastern Texas. The 2006 program extended the earlier study to address climate-air quality linkages and probe nighttime NOx and NOy chemistry.

The Bay Region Atmospheric Chemistry Experiment (BRACE) was conceived in response to persistent increasing trends of nitrogen oxide emissions in Florida, in order to assess potential effects on air quality and the ecological health of Tampa Bay and its surroundings. BRACE began in 2000 and has included both long-term and short-term intensive measurement campaigns, focusing on assessment of atmospheric nitrogen deposition to Tampa Bay. This program helped develop modeling approaches for numerous eastern U.S. estuaries, and has been augmented by additional field campaigns elsewhere, most notably in eastern North Carolina. Key participants included the Florida Department of Environmental Protection, Tampa Bay Electric Company, EPA, NOAA, Argonne National Laboratory, numerous universities, and several additional Florida state agencies.

The PM Supersites program complemented routine $PM_{2.5}$ monitoring by deploying research instrumentation in intensive field campaigns to obtain highly time-resolved data on multiple aerosol physical and chemical properties in major U.S. cities (Atlanta, Baltimore, Fresno, Houston, Los Angeles, Pittsburgh, St. Louis, and New York).

These data sets, spanning portions of 1999 – 2004 with some sites operating in later years, addressed three primary objectives: development of monitoring methods and transfer to operational agencies; support for health effects research; and State Implementation Plan (SIP) development. Several findings are synthesized in dedicated special journal issues (Pandis et al., 2005; Geller and Solomon, 2006; Solomon et al., 2008).

The Los Angeles Supersite and the Southern California Particle Matter Center spawned interest in near-roadway characterizations by providing measurements of particles near highways, particularly ultrafine particle numbers. This work showed very high particle concentrations near the highway (Zhu et al., 2002), with concentrations decreasing and size distributions changing with increased distance from the roadway. Near-roadway studies have since been undertaken, primarily by EPA, in several other cities, including Detroit, Raleigh, and Las Vegas. Near-roadway measurements and studies provide an important bridge between atmospheric and health sciences, given the combination of highly variable pollutant concentrations in populated locations. The EPA is developing technical requirements for monitoring to improve the understanding of exposures in near roadway environments.

The Polar Study using Aircraft, Remote Sensing, Surface Measurements and Models of Climate Chemistry, Aerosols, and Transport (POLARCAT) was a coordinated international series of field studies conducted as part of the International Polar Year 2007-2008 (IPY).

Under this activity, NASA led the 2008 Arctic Research of the Composition of the Troposphere from Aircraft and Satellites (ARCTAS) project, largely based in Canada. This field study addressed a variety of issues impacting the Arctic atmosphere, including Eurasian and North American fires, halogen chemistry, light absorbing carbon, and persistent pollutants. Prior to ARCTAS deployment in Canada, NASA had conducted a series of flights with scientists from the California Air Resources Board (CARB). These flights, already configured for the ARCTAS mission, examined California's atmosphere to better understand the chemical dynamics of smog and greenhouse gases over the state.

NOAA also led two field studies as part of the IPY: Aerosol, Radiation, and Cloud Processes affecting Arctic Climate (ARCPAC) and International Chemistry Experiment in the Arctic LOwer Troposphere (ICEALOT). ARCPAC involved the deployment of the NOAA WP-3D aircraft in Alaska, and ICEALOT involved the deployment of the Woods Hole Research Vessel *Knorr* in the North Atlantic. Additionally, DOE led the Indirect and Semi-Direct Aerosol Campaign (ISDAC) as part of the IPY.

The CalNex campaign took place during the spring/summer of 2010, and was designed to build on existing California programs that address air quality and climate linkages. Under NASA's Earth Venture missions in the Earth Science System Pathfinder program, the Deriving Information on Surface Conditions from Column and Vertically Resolved Observations Relevant to Air Quality (DISCOVER-AQ) mission is conducting a series of field missions.

The overarching objective is to use targeted airborne and ground-based observations to improve the interpretation of satellite observations in diagnosing near-surface conditions relating to air quality. The first of the field studies was conducted over the greater metropolitan area of Baltimore, MD, with a complex multi-platform observing system, providing multiple perspectives on the factors that control air quality and influence the ability to monitor pollution events from space. Subsequent deployments were conducted in California's San Joaquin Valley in January 2013 and Houston, Texas, in September 2013. The DISCOVER-AQ mission will conduct a final deployment in Denver, Colorado, in summer 2014.

Field Campaign Websites

ARCTAS:	http://www.espo.nasa.gov/arctas/
CalNex:	http://www.esrl.noaa.gov/csd/calnex/
DISCOVER-AQ:	http://discover-aq.larc.nasa.gov
ICARTT:	http://www.esrl.noaa.gov/csd/ICARTT/index.shtml
INTEX-B:	http://www.espo.nasa.gov/intex-b/index.html
INTEX-NA:	http://cloud1.arc.nasa.gov/intex-na/
MILAGRO:	http://www.eol.ucar.edu/projects/milagro/
NEAQS - ITCT 2004:	http://www.esrl.noaa.gov/csd/2004/
POLARCAT:	http://www.nilu.no/Portals/0/IMG/Forskning/NILU-Polaraaret-A4-web- NY.pdf
TRACE-P:	http://www-air.larc.nasa.gov/missions/tracep/traccp.htm
TexAQS & TexAQS II:	http://www.esrl.noaa.gov/csd/2006/

Satellite–Based Air Quality Observations

An extensive array of satellite-based systems (Appendix H) has been established by the United States and European Union countries to measure total atmospheric columns and limited vertical profiles of several key species. In the United States, these programs are led by the National Aeronautics and Space Administration (NASA) and the National Oceanic and Atmospheric Administration (NOAA). Non-government organizations have also played significant roles, such as the Harvard-Smithsonian Astrophysical Laboratory and the National Center for Atmospheric Research (NCAR). In Europe, the satellite measurements relevant to air quality are part of the large Global Monitoring for Environmental and Security (GMES) program, now renamed Copernicus. The satellite component is led by the European Space Agency (ESA) and the European Organization for the Exploitation of Meteorological Satellites (EUMETSAT). A number of other national programs, beyond the U.S. and European programs, provide relevant measurements (listed in Appendix H). NASA and ESA typically demonstrate new capabilities for Earth observations, while NOAA and EUMETSAT conduct long-term operational observations. Under U.S. programs, a suite of satellites including Terra, Aqua, Aura, CALIPSO (Cloud-Aerosol Lidar and Infrared Pathfinder Satellite Observation), as well as NOAA-17, NOAA-18, NOAA-19, and

Suomi NPP (National Polar-orbiting Partnership), have been launched in the last two decades. Collectively, they measure columns and/or profiles of aerosol optical depth (AOD), O_3, H_2O, CO, CH_4, SO_2, NO_2, CFCs, other pollutants, and atmospheric parameters such as temperature, cloud properties, and water vapor. Most of these satellites have a near-polar low Earth orbit (LEO), passing twice per day over a given location. For many species, measurements are only possible during daylight, so only one measurement is made per day per instrument.

The Earth Observing System (EOS) Afternoon Constellation, or "A-Train," is a group of several of these satellites (Aqua, Aura, CALIPSO, CloudSat, and (previously) PARASOL (Polarization & Anisotropy of Reflectances for Atmospheric Sciences coupled with Observations from a Lidar) that fly in formation, crossing the equator a few minutes apart near 1:30 PM local time. The near-simultaneous observations from these satellites produce a rich picture of earth weather, climate, and atmospheric conditions. The OCO (Orbiting Carbon Observatory) mission, which suffered a launch vehicle failure in early 2009, was to join the A-Train to measure CO_2 with the precision required to map global distribution of CO_2 sources and sinks on regional scales. A U.S. replacement mission for OCO is scheduled for a July 2014 launch. In the meantime, Japan's Greenhouse gases Observing Satellite (GOSAT), launched in January, 2009, is globally monitoring CO_2, CH_4, and aerosols.

NOAA's National Environmental Satellite and Data Information Service (NESDIS) oversees operations of U.S. Geostationary and Polar-orbiting Operational Environmental Satellite programs (GOES and POES), providing imagery for weather forecasting and observations of light scattering relevant to aerosol characterizations. Under the GOES program, NOAA is scheduled to launch in 2016 the first of the GOES-R series of next generation geostationary weather satellites with the Advanced Baseline Imager (ABI). The ABI will provide aerosol optical depth measurements at a spatial resolution of 2 km every 5 minutes over the continental U.S. and Canada and, every 15 minutes, the entire disk of the northern and southern hemispheres. The Suomi National Polar-orbiting Partnership (NPP) satellite launched on October 28, 2011, with sensors such as OMPS (Ozone Mapping and Profiler Suite), VIIRS (Visible/Infrared Imaging Radiometer Suite), and CrIS (Cross-track Infrared Sounder), provides aerosol and trace gas information at much higher spatial resolutions than previously possible. For example, VIIRS aerosol optical depth and suspended matter (aerosol type) products are available at 750 m resolution. The joint NASA-NOAA Suomi NPP mission serves as a proving ground for the series of NOAA Joint Polar Satellite System (JPSS) operational satellites to be flown in the next two decades.

Beginning in 1995, ESA and EUMETSAT have flown a series of LEO satellites that typically provide measurements of these same air quality relevant species at approximately 9:30 AM local time (Ingmann et al., 2012). Europe is planning to continue these observations through the mid-2020's with the operational MetOp satellite series. Europe is also planning to continue and improve the data records available in the 1:30 PM orbit (currently provided from NASA Aura) by launching the Sentinel-5 precursor satellite (Veefkind at al., 2012) and Sentinel-5 missions in 2015 and 2022 respectively.

For the future, NASA and partner agencies are studying additional satellite platforms capable of measuring trace gases and aerosols to enhance the characterization of tropospheric air quality from space (NRC, 2007; Fishman et al., 2008; Fishman et al., 2012). The National Research Council (NRC) has recommended that NASA implement a number of "Decadal Survey Missions" over the next decade, in addition to implementing the JPSS and GOES programs.

These "Decadal Survey Missions" include the following:

(1) The Geostationary Coastal and Air Pollution Events (GEO-CAPE) will partially focus on supporting air quality assessments and forecasts by measuring atmospheric columns with a frequency of one hour from a geostationary space platform.
(2) The Active Sensing of CO_2 Emissions over Nights, Days, and Seasons (ASCENDS) mission will produce global atmospheric column CO_2 measurements without seasonal, latitudinal, or diurnal bias using simultaneous laser remote sensing.
(3) The Aerosols, Clouds and Ecosystems (ACE) mission will consist of a lidar for characterizing aerosol height and properties, and a polarimeter for determining aerosol types.
(4) The Global Atmospheric Composition Mission (GACM) will focus on ozone and related gases for intercontinental air quality and stratospheric ozone layer monitoring, from a LEO space platform.

Several nations (including the United States and countries in Europe and Asia) are currently planning to launch geostationary satellites capable of providing a common suite of air quality relevant measurements (O_3, NO_2, HCHO, aerosol, CO, SO_2, and other pollutant species) at hourly temporal resolution throughout daylight hours and at 5-10 km horizontal resolution

(Committee on Earth Observation Satellites [CEOS], 2011). Geostationary Earth Orbit (GEO) satellites provide continuous observations over one part of the globe. A constellation of at least three GEO satellites can together provide hourly near-global observations of populated regions. Such observations will bring revolutionary new capability for satellite monitoring of emissions and receptor processes (Lahoz et al., 2012). Under the European GMES (Copernicus) program, ESA will launch the Sentinel-4 platform in 2019. The East Asia mission, Geo-KOMPSAT (Geostationary Korea Multi-Purpose Satellite) is led by the Korea Aerospace Research Institute (KARI) and scheduled to launch in 2018. GEO-CAPE is the NASA Decadal Survey mission under consideration in this category. In November 2012, NASA selected the Tropospheric Emissions: Monitoring of Pollution (TEMPO) proposal as the first investigation in the new Earth Venture-Instrument mission. TEMPO provides much, but not all, of the capability planned for GEO-CAPE atmospheric observations. The TEMPO instrument will be delivered in 2017 for an anticipated 2018 launch on a geostationary host satellite. TEMPO will be placed in an orbit location from which it can observe much of North America. TEMPO and the ESA and KARI missions will each provide hourly measurements of ozone, aerosols, and their precursors, such as NO_2, SO_2, and HCHO over their respective regions, together providing unprecedented highly space- and time- resolved observations of the Northern Hemisphere.

Satellite data complement surface networks and aircraft campaigns, and are essential tools for evaluating models and improving emissions inventories.

There has been inter-agency cooperation between the National Institute of Environmental Health Sciences (NIEHS), EPA, NASA, and NOAA in exploring the use of satellite observations to support epidemiologic studies. Satellite observations do not directly correspond to in situ measurements of pollutant concentrations. Thus, the use of satellite data for air quality forecasting, management, health effects studies, and climate change assessments is complex and involves integration of models and surface observations. While satellites offer global or near-global coverage of several important species, there are basic limitations in using a space platform to effectively probe the lower levels of the atmosphere where pollution exposures occur. Understanding these limitations is important for gauging how these systems can best complement ground- based networks and support air quality management assessments.

Attributes of Air Quality Satellite Data Products

Fundamental Limitations

Most satellite air quality observations are based on spectroscopic techniques using reflected, scattered, or emitted solar radiation as a broad source of radiation. Although the science of measuring trace gases and aerosols from space is relatively mature, interference related to variable surface reflectivity, cloud attenuation, and overlapping spectra of other chemical species require significant data processing and treatment. Even with this sophisticated data retrieval, data products will contain spatial gaps for a given time period due to cloud interference and other issues, such as sun glint, etc. This is particularly true for instruments in a polar orbit, where the instrument may see an area only one time during the daytime. For example, aerosol events occurring at the same time as clouds are often screened as a result of cloud fractions being too high to produce an accurate aerosol optical depth (AOD) product for NASA's Moderate Resolution Imaging Spectroradiometers (MODIS) aboard the Aqua and Terra satellites.

Most satellite sensors sum over the entire column of air from satellite to ground, providing total column densities of trace gases or aerosols, whereas concentrations near the surface are used to define air quality. Some information on the vertical distribution of certain species can be obtained by using multiple observing angles for instruments in LEO, by limb sounding, active sensing, or other methods. For example, CALIOP aboard the CALIPSO satellite resolves aerosol vertical distributions every 30 meters. It is important to note that, for certain important trace gases (e.g., NO_2, SO_2, and HCHO) and aerosols, the majority of mass resides in the boundary layer of the lower troposphere, enabling associations linking column data to surface concentrations or emissions fields. In the eastern United States, reasonable correlations (Engel-Cox et al., 2004) have been developed between concentrations from ground level $PM_{2.5}$ stations and MODIS AOD using a fixed relationship. In the western United States, correlations are poor due to excessive surface light scattering from the relatively barren land surfaces. To better account for aerosols aloft, output from a chemical transport model has been applied to develop concentrations of $PM_{2.5}$ from MISR AOD (Liu et al., 2004).

In contrast to aerosols, most ozone resides in the stratosphere. Various techniques have been developed to extract the stratospheric signal to yield a tropospheric ozone residual (TOR), based on known homogeneities in the stratosphere and the use of chemical transport models and multiple measurements. Early approaches (Fishman, 1978), before and during the Total Ozone Mapping Spectrometer (TOMS) missions, combined limb (angled view to characterize the stratosphere) and nadir (downward view, characterizing the total column) techniques to derive tropospheric ozone residuals. The 2004 launch of NASA's Aura mission, with multiple ozone sensors, is starting to produce more refined tropospheric ozone maps. For example, direct derivation of tropospheric column ozone is possible from the Ozone Monitoring Instrument (OMI) on Aura (Liu et al., 2009). However, differentiating ozone in the boundary layer from that in the free troposphere continues to pose significant challenges.

This difficulty stems from strong molecular scattering of UV radiation in the boundary layer, and surface emission in the thermal IR range. Notably, the geostationary TEMPO instrument will make ozone measurements in both UV and visible (Chappuis-band) wavelengths to provide improved sensitivity to ozone in the boundary layer using multispectral retrievals.

Temporal Coverage

The near polar orbiting tracks of most LEO satellites deliver at most twice daily snapshot measurements of trace gas species (approximately 12 hours apart). Measurements of many species can only be taken once during the single daytime overpass. Consequently, these instruments can only observe temporal patterns of pollutants or time-integrations of pollutant concentrations at daily or longer scales. Furthermore, instruments in LEO have only a short exposure to each Earth scene, limiting the signal-to-noise ratio. For many LEO products, observations for a given day are quite noisy, and weekly or monthly averages are more typically used. Geostationary (GEO) satellite platforms, such as the NOAA GOES systems (http://www.oso.noaa.gov/goes/index.htm), do provide near-continuous monitoring of physical parameters for weather tracking and forecasting purposes. The ability of a GEO instrument to observe an area for longer time periods potentially enables a sufficient signal-to-noise ratio to make short time period observations meaningful, on the order of one hour. TEMPO will provide the first-ever observations of tropospheric ozone and its precursors over North America from geostationary orbit.

Spatial Coverage

Polar orbiting satellites typically provide horizontal spatial resolution between 10 and 100 km for atmospheric composition. Spatial resolution less than 10 km is possible with GEO and LEO platforms. Satellite observations of pollutants above the surface complement ground-based in-situ measurement networks – especially considering that a considerable fraction of pollutant mass resides well above the Earth's surface. As noted above, the sensitivity of satellites to pollutants at elevated heights can obscure measurements of the boundary layer, and, in general, satellite data products contain little or no information about the vertical distribution of pollutants.

Furthermore, one technique used to obtain vertical distribution information, the comparison of nadir and limb observations, is usable from GEO only if complementary limb observations are available from one or more polar orbiting satellites. The limitations of satellite data are further enumerated and elaborated (Vijayaraghavan et al., 2007) for the Global Ozone Monitoring Experiments (GOME) sensors for O_3, NO_2, and HCHO.

Measuring near-surface pollution is one of the most challenging problems for Earth observations from space. The NASA DISCOVER-AQ missions (2011-2014) are expected to help address this issue, along with some of the fundamental limitations discussed above. These missions are anticipated to help bridge the existing knowledge gap limiting the relevance of satellite observations to air quality. At the same time, they will provide insights into the potential benefits of higher time and space resolved measurements, as anticipated by GEO platforms.

Current Use of Satellite Data in Air Quality Management

In broad terms, satellite measurements serve as complements to other surface-based and aircraft measurement programs and air quality models. Satellite applications for air quality forecasting and assessments are covered extensively in the published literature (Martin, 2008; Fishman et al., 2008; Vijayaraghavan et al., 2007). The following summary describes how these data are most effectively incorporated in air quality assessments, albeit not capturing the full breadth of applications of satellite observations. Four general methods are applied in the use of satellite observations for air quality assessments:

Detecting Evidence of Long-Range Transport

Satellite data support assessments of air quality on hemispheric and global scales, and assessments of long-range transport. These are projected to be of increasing importance to North American air quality management. Trans-oceanic air pollution transport can be observed with satellites, and direct observational evidence of this phenomenon has been clearly visible in satellite imagery (Figure 3).

Characterizing Emissions and Air Quality Model Support

Satellite observations play an important role in emissions characterization, particularly for source regions and sectors that have inadequate bottom-up inventories.

Applications include improving inventories from poorly-characterized, developing regions in Asia, and from "natural" sources such as lightning and soil NOx, biogenically emitted VOCs, wildfire particulate matter emissions, and agricultural based ammonia/NH_3.

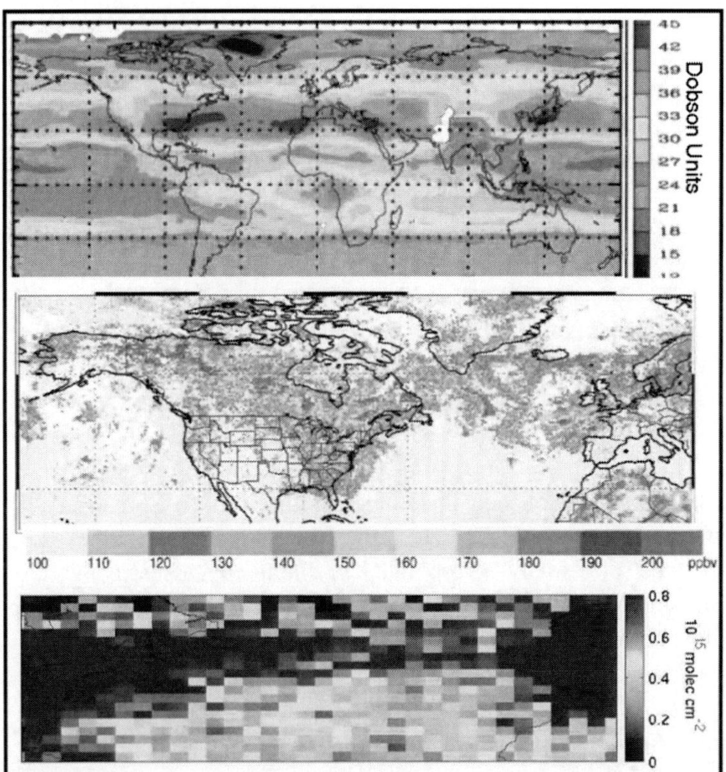

Figure 3. Panels capturing trans-Atlantic transport. Top: summer 1997 tropospheric ozone from GOME (Liu et al., 2006). Middle: CO column totals from MOPITT for July 2004 (Pfister et al., 2005). Bottom: Tropospheric NO_2 from SCIAMACHY for summer 2004 (Martin et al., 2006).

The location and source strength of wildfire plumes detected from space serve as important inputs for annual emissions inventories (Martin et al., 2006; Soja et al., 2009) used in EPA air quality models, such as the Community Multiscale Air Quality Model (CMAQ), and processed as part of EPA's National Emissions Inventory. Satellite-based NO_2 (Wang et al., 2010) and SO_2 (Lu et al., 2010) data have illustrated the dramatic increases in Asian NOx and SOx emissions over the last decade. Increasing NOx and SOx emissions associated with gas and oil extraction from 2005 - 2010 in the Canadian oil sands have been identified by satellite NO_2 and SO_2 observations (McLinden et al., 2012). Biogenic VOC emission estimates have also been developed (Millet et al., 2008) using satellite measurements of formaldehyde, an oxidized product of directly-emitted isoprene. Advances in processing NH_3 signals from the Tropospheric Emission Spectrometer (TES) have led to improved spatial and temporal characterization of NH_3 surface patterns (Pinder et al., 2011), affording a potentially valuable tool to improve emissions estimates and evaluate models for sparsely measured air pollutants at the surface.

Air quality management depends on models for complex environmental characterizations that cannot be achieved through observations alone. Satellite-based enhancements to surface monitoring networks support the evaluation of these models by inter-comparison and improving emission inventories. Inverse modeling (the process of using a chemistry transport model (CTM) to estimate the emissions that reproduce the satellite observations) is frequently used in deriving "top-down" emissions that can improve and update emissions estimates from "bottom-up" inventories. Satellite observations of tropospheric ozone can be used as the basis for boundary condition inputs to regional air quality models. Satellite (OMI, GOME, and GOME-2) observations of NO_2 and HCHO can also be used to diagnose the sensitivity of ozone production to NOx or VOCs (Martin et al., 2004; Duncan et al., 2010).

Sustained long-term satellite observations support accountability analyses of the efficacy of environmental regulations. Due to a lack of surface-based true NO_2 measurements, satellite observations have been the most useful indicators of progress in the NOx State Implementation Plan (SIP) Call for Revision (Figure 4). Similarly, substantial declines in SO_2 emissions from coal power production facilities in the Eastern United States from 2005–2010, implemented through various cap and trade rules for major emission sources, are well correlated with satellite-derived, observed, SO_2 trends (Fioletov et al., 2011). A particular advantage of satellite observations is their ability to provide an abundance of observations with the same methodology across all locations on the planet. Consequently, global views emerge of pollutant trends capturing United States and European progress in emissions abatement, contrasted with emissions growth due to industrial expansion in Asia and elsewhere (Figure 5).

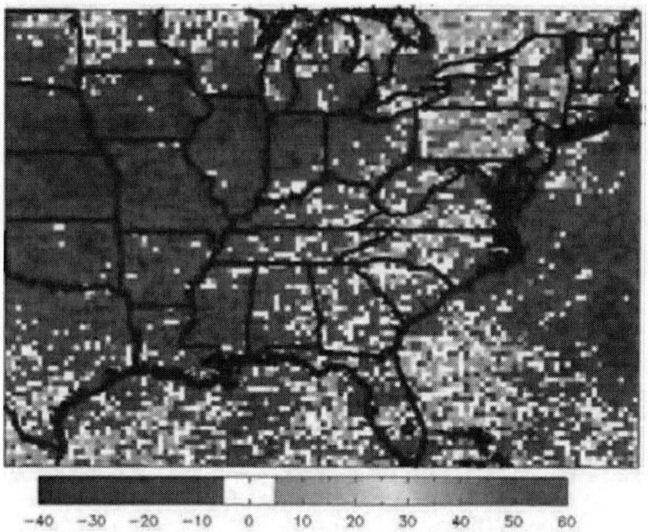

Figure 4. Percentage difference (July 2008 minus July 2005) in satellite-derived tropospheric NO_2 column amounts. Satellite data provide measurements that can be compared over large domains, suitable for trends analysis (Neil et al., 2009).

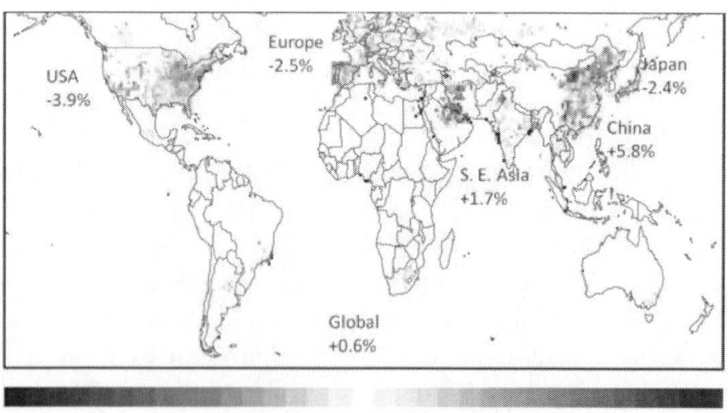

Figure 5. Percentage differences (2010 – 2005) in satellite-derived tropospheric NO_2 column amounts based on OMI observations following the methodology of Lamsal et al., (2011).

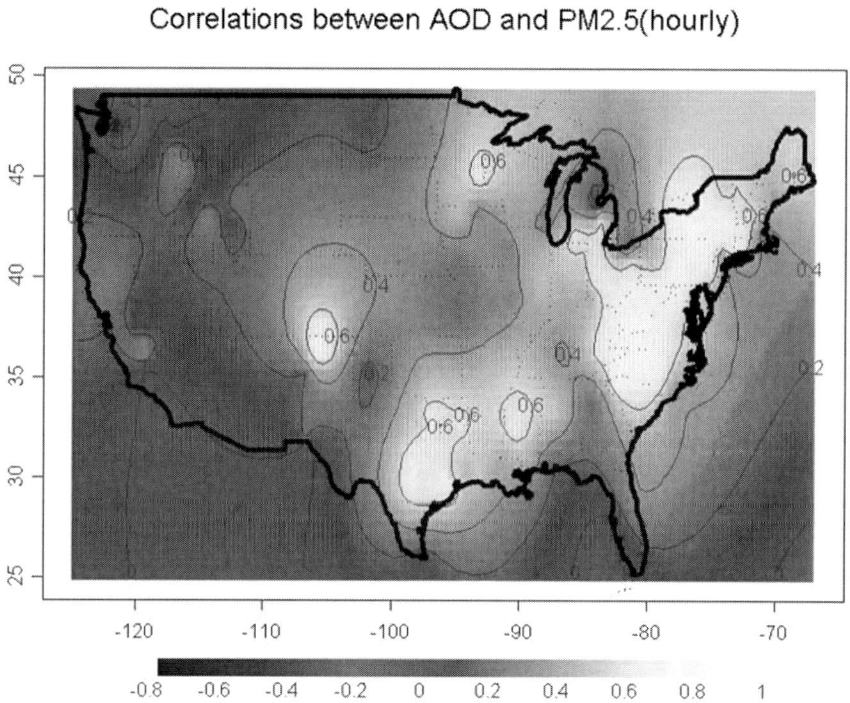

Figure 6. Correlation Coefficient between MODIS Aerosol Optical Depth (AOD) and hourly $PM_{2.5}$ surface sites from April - September, 2002 (Engel-Cox et al., 2004).

These success stories of infusing satellite observations into air quality assessments are notable as many of these "natural" and international emission source areas constitute a substantial fraction of atmospheric mass loadings that historically were generated without the benefit of an observation based evaluation component.

Serving as Surrogates to Filling Gaps in Surface Networks

Satellite observations are assisting the air quality community by providing data that cover broad spatial areas lacking ground-based monitors and, more importantly, a vertical (or column) complement to surface-based networks (Al-Saadi et al., 2005). Although 'breathing- zone' monitoring is essential, most pollutant mass resides outside the domain of surface stations. Pollutant levels aloft often correlate well with surface conditions during well- mixed afternoon conditions in stable pressure systems. This offers the potential for "gap filling" in the surface-based networks (Figure 6). However, the appeal of satellite observations to fill gaps in surface measurement must be tempered by the limitations of using space-based measurements to characterize near surface conditions.

Providing Support During Episodes and Forecasting Air Quality

Fire and dust events can produce atypically bad air quality in areas that do not usually experience poor air quality. Satellite data have played an important role in "exceptional event" analyses in determining attainment of the National Ambient Air Quality Standards. Satellite data can act as a surrogate for gaps in surface monitoring data, and the imagery can have an important public information role. Air quality forecast models rely on satellite data for verification (Al-Saadi et al., 2005; Kondragunta et al., 2008), offering the potential for assimilating aerosol and ozone gaseous precursor satellite data to constrain model predictions in field studies and next day air quality (Mathur, 2008; Pierce et al., 2009).

Observation Programs for Climate, Background Concentrations, Stratospheric Ozone, and Long-Range Pollutant Transport

NOAA and NASA are the lead Federal agencies for a variety of observation programs focused on climate change, background concentrations of trace gases in areas free of large local sources, stratospheric ozone, and pollutant transport. These data systems and networks include surface measurements, vertical profiling, and measurements of atmospheric columns (Appendix I). The Department of Energy (DOE) is also engaged in observation programs addressing climate change. Many of these observation programs rely on partnerships across U.S. Federal agencies and collaborations with international organizations, such as the World Meteorological Organization (WMO).

This section focuses on observation networks in the United States, which are largely the responsibility of NOAA, NASA, and DOE, with various levels of participation from partner agencies. The assortment of gases and aerosols that are important climate moderators and/or key air quality indicator and precursor species include: CO_2, N_2O, H_2O, CH_4, O_3, CO, aerosols, and halogenated compounds, including CFC replacements.

Greenhouse Gas (GHG) Observation Systems

NOAA, NASA, and DOE are lead Federal agencies for several programs addressing greenhouse gas trends, sources, sinks, and fluxes, many in partnership with each other and with international organizations. Although these programs are focused on carbon dioxide budgets, other important GHGs are included where feasible. In particular, methane (CH_4) acts as a strong climate forcing gas and influences global background levels of tropospheric ozone (West et al., 2007). The major GHG observation programs with sites in the United States include:

NOAA Global Cooperative Air Sampling Network

This cooperative network includes over 100 remotely located surface stations worldwide, and a series of ship routes. Weekly samples from this network are used to determine global CO_2, N_2O, CH_4, and CFC concentration trends.

AmeriFlux Network

DOE coordinates a multi-Federal agency group (with NOAA, USDA, NSF) overseeing the AmeriFlux network of ~90 active sites in the United States.

The network sites are largely micrometeorological towers ranging from a few to hundreds of meters in height. Each tower is instrumented with a fast CO_2 monitor and wind sensors, allowing calculation of the flux of CO_2 between the surface, vegetation, and the atmosphere. AmeriFlux is a component of the worldwide Fluxnet system of CO_2 flux networks, tracking storage of carbon in terrestrial systems (http://www.fluxnet.ornl.gov/fluxnet/index.cfm). The emerging NEON (National Ecological Observatory Network; http://www.neoninc.org/science/domains) is an NSF supported program that will provide 62 sites across the U.S., some of which are co-located with AMERIFLUX sites.

Vertical Profile, Atmospheric Column, and Satellite Observations

The NOAA network of 8 tall towers (100 – 500 m) provides regionally representative, near-continuous, boundary layer measurements of CO_2 and related gases. As noted in Section 2, NASA's Orbiting Carbon Observatory (OCO) satellite mission was intended to be the Nation's primary remote sensing platform for CO_2, providing a continental and oceanic scale complement to ground based systems and aircraft programs. In light of OCO's

2009 launch failure, and delay of a replacement mission, climate monitoring relies for the present on Japan's Greenhouse Gases Observing Satellite (GOSAT). GOSAT was launched in January 2009, and measures CO_2 and CH_4 globally. Several United States and European satellite sensors (Aqua AIRS, NPP CrIS, and MetOp-1 IASI) also provide some profiling capabilities of greenhouse gases (O_3, CH_4, and CO_2) globally.

Monitoring of Background Air Quality and Long-Range Transport

Network of Remote (Sentinel) Surface Observation Stations

Remote surface stations, located in areas relatively free from nearby sources, can characterize background pollutant levels, transport on regional and hemispheric scales, and boundary conditions for air quality models. NOAA maintains six baseline sites or surface "sentinel" stations designed to capture long-term trends and atmospheric background air pollutant concentrations (Mauna Loa, HI; Trinidad Head, CA; Barrow, AK; American Samoa; South Pole; and Greenland). These are part of a worldwide Global Atmospheric Watch (GAW) network of baseline sites coordinated by the WMO (Figure 7).

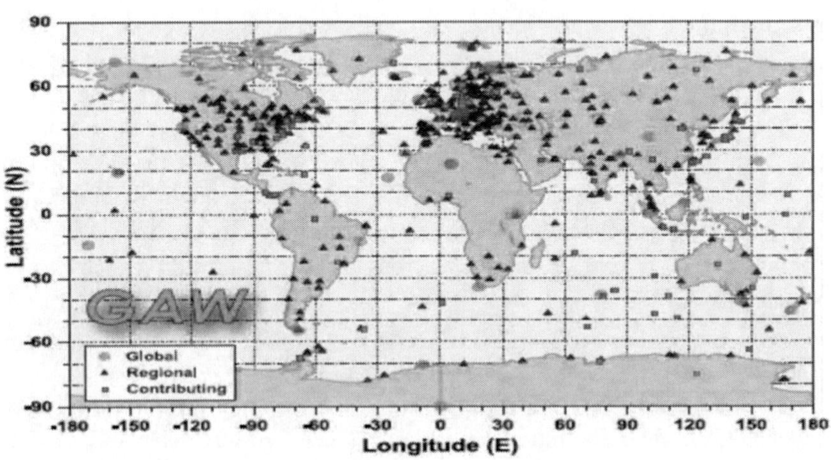

Figure 7. Network of surface-based remote observatories organized through the World Meteorological Organization's Global Atmospheric Watch (GAW).

NASA Fixed Site Observation Networks

The AGAGE (Advanced Global Atmospheric Gases Experiment; http://agage.eas.gatech.edu/) and its predecessors (the Atmospheric Life Experiment, ALE, and the Global Atmospheric Gases Experiment, GAGE) monitor a variety of climate forcing gases, CFCs, and reactive trace gases at remote "sentinel" sites throughout the world.

Monitoring of Pollutants Aloft: Profiles and Total Columns

Vertical profiling and total atmospheric column measurements provide important complements to near surface observations. Observations aloft provide insight into background levels and transport phenomena, and are key metrics for model evaluation. It is difficult to rely on surface-based measurements to characterize these higher elevations because of complex near-surface deposition, removal, and micrometeorological processes. Programs to improve vertical profiles include a variety of aircraft, sondes (self-contained signal transmitters), remote sensing, tall towers, and special field programs.

These are largely managed by NOAA, NASA, NSF, and DOE. Proper siting and measurement techniques can produce observations that support assessments of climate change, stratospheric ozone, baseline concentrations, and long range and regional transport. These approaches include sampling throughout the atmospheric column (for total column and vertical profiles) and, particularly for fixed surface observations, measurements in locations that are relatively source free. Although climate and stratospheric ozone depletion assessments benefit from characterizing the full atmospheric column through the stratosphere (~35 km), systems designed to capture the entire atmospheric column often have insufficient resolution or precision in the boundary layer (~5 km) for surface-oriented air quality assessments.

NOAA Surface and Aircraft Based Air Quality Measurement Programs

NOAA conducts a variety of routinely scheduled fixed-site and aircraft-based measurement programs, along with a series of intensive special field campaigns, to provide observations addressing a variety of climate, stratospheric ozone depletion, and planetary boundary layer air quality issues. These programs are a source of data on conditions aloft. Core elements of these measurement programs include:

- An ozone radiosonde network (8 sites, 4 in the United States) providing one day per week vertical ozone profiles with approximately 100m resolution from the surface through the stratosphere;
- The Dobson ozone spectrometer network (16 station cooperative network, 11 in the United States) providing near continuous daytime total atmospheric column ozone data;
- Routine aircraft flights that characterize the vertical distribution of air pollutant species (O_3, CO, CH_4, CO_2, N_2O, SF_6) for climate and air quality assessments;
- Tall tower sites (8) that are part of the larger interagency North American Carbon Program (NACP), designed to characterize carbon sources, sinks, and removal processes. These towers are currently located throughout the continental United States using television and cell phone towers (100 – 500 m tall). They provide near-continuous regionally representative boundary layer measurements of CO_2, CO, CH_4, and associated fluxes, various trace gases, and meteorological parameters (http://www.esrl.noaa.gov/gmd/ccgg/towers/ index.html);
- Special intensive studies, often in collaboration with NASA, with aircraft focusing on regional U.S. air quality issues, typically conducted every two years (Section 2) and for satellite validation.

These programs, in combination with NOAA remote surface-based measurement observatories (http://www.esrl.noaa.gov/gmd/about/ airquality. html), provide long term records of baseline air quality from the surface through the stratosphere. They represent a substantial component of the United States' contribution to international monitoring, much of which is organized through WMO's GAW program.

European-Based Aircraft Programs

Two programs, MOZAIC (Measurement of OZone and water vapour by AIrbus in-service aircraft; operated since 1994) and CARIBIC (Civil Aircraft for the Regular Investigation of the Atmosphere Based on an Instrument Container; operated since 2004) measure air quality parameters from in-service, scheduled passenger aircraft based in Europe. These programs provide widely distributed, frequent, measurements in the upper troposphere, including over the Atlantic.

On takeoff and landing, they provide vertical profiles over cities, including cities in North America. These programs provide the most extensive, routinely-collected, vertically- distributed air quality data from throughout the troposphere.

MOZAIC: http://www.iagos.fr/web/rubrique3.html

CARIBIC: http://www.caribic-atmospheric.com/

Aircraft based observation programs are transitioning to the newly initiated IAGOS (In- service Aircraft for the Global Observing System) program, a partnership between European research institutions, universities, Airbus Industries, and commercial airlines around the world, including the United States. The purpose of IAGOS is to establish and operate a distributed infrastructure for long-term observations of atmospheric trace gases (O_3, CO, CO_2, CH_4, NOy, NOx, H_2O), particulate matter, and cloud droplet backscatter on a global scale. The initial fleet comprises 10 - 20 long-range in-service aircraft belonging to airlines based throughout the world. The data will be relevant for studies of air quality, long-range transport of pollution plumes, climate change, and the impacts of aviation on the composition of the atmosphere. The data will also be made available in near real-time for use by air quality forecast models, and in air quality modeling for environmental policy-making.

Satellite Air Quality Validation Programs

NASA oversees the operation of several programs monitoring the lower atmosphere, designed to complement and evaluate satellite products described in Section 2. These programs generally have a broader hemispheric- or global-scale perspective, which often overlaps with regionally-focused initiatives.

Network for the Detection of Atmospheric Composition Change (NDACC)

The NDACC is an international activity focused on obtaining high quality measurements of a broad range of atmospheric chemical species and parameters. This network includes more than 70 remote-sensing research stations. Originally focused on the stratosphere, with an emphasis on the ozone layer, the scope of the NDACC has expanded to cover both the stratosphere and troposphere.

A variety of working groups operate under NDACC, each focused on a particular measurement or technique, including Dobson/Brewer, FTIR spectrometers, lidar, microwave radiometers, satellite measurements, sondes, UV/Vis spectrometers, spectral UV, and water vapor. This network has been in operation since 1991 (http://www.ndsc.ncep.noaa.gov/).

Lidar Networks

Lidar, analogous to radar, uses backscattered laser light to profile aerosols, gas-phase species, or other parameters, such as temperature, above a site. In addition to use during individual field studies, there are three fixed-site, long-term lidar monitoring networks in the United States.

The Micro Pulse Lidar Network (MPLNET) is coordinated by NASA, and operates from 14 stations (4 in the United States). MPLNET profiles aerosols, and most sensors are co- located with AERONET (see below). The NOAA CREST lidar network comprises four sensors that profile aerosols, and are operated by academic institutions in the eastern United States. Three of these sites can also profile water vapor. The international Network for the Detection of Atmospheric Composition Change (NDACC, described above) includes 17 lidars (3 in the United States). These networks are affiliated with a number of networks operating overseas in the WMO/GAW Aerosol Lidar Observation Network (GALION).

Selected Meteorological Observation Systems

Two categories of above-surface meteorological systems are included here because of their linkage to air quality assessments. First, solar radiation networks provide estimates of atmospheric aerosols and various trace gases, in addition to basic data for the radiation components of models. Second, systems that enable estimation of the height of the planetary boundary layer are important for near-surface air quality analyses and model applications.

Solar Radiation Networks

Full spectrum and wavelength-specific solar radiation measurements provide data used to characterize energy budgets for meteorological models, climate change assessments, and atmospheric column aerosol light scattering. They also serve as direct indicators of UV radiation exposure, relevant to human and ecosystem health and agriculture. A variety of Federal agencies have participated in measurement programs, including NOAA, NASA, EPA, USDA, and the National Park Service.

The Aerosol Robotic Network (AERONET) is a NASA-organized, collaborative, global network of sun photometers providing ground-based aerosol optical depth (AOD) estimates used primarily to evaluate satellite aerosol measurements. NOAA's Surface Radiation Budget Network (SURFRAD) is part of the global Baseline Surface Radiation Network

(BSRN). It is an important surface complement to satellites, and is used to characterize surface energy balances and support a variety of global scale climate models.

The Brewer UV spectrophotometer networks started in 1994 with EPA's UVNet program. UVNet included over 20 sites until funding ended in 2004. A subset of six sites supported by EPA and NOAA is operating as the NOAA-EPA Brewer Spectrophotometer UV and Ozone Network (NEUBrew). Brewer networks are designed to monitor UV radiation at the surface to understand effects on human and ecosystem health and agriculture. EPA has provided funding to study the relationship between changes in stratospheric ozone and UV at the surface. The Brewer instruments are capable of providing total column ozone and SO_2 estimates.

Observations for Evaluating PBL Heights

Planetary boundary layer (PBL) height (or mixed layer height) is an important physical parameter in air quality models. PBL is a derived quantity based largely on vertical temperature profiles and refractive index structure parameters. The deployment of the NOAA Profiler Network (NPN: http://www.profiler.noaa.gov/npn/) over the last decade has added a near-continuous stream of wind vector data to complement the National Weather Service's (NWS) rawinsonde (radio tracked sonde) network, which provides twice daily soundings spread across nearly 100 locations throughout the United States. NPN consists of 35 unmanned Doppler radar sites profiling the troposphere, concentrated in the central United States and designed for forecasting violent weather. The PAMS program supports ~20 radar profilers that provide highly resolved wind profiles of the boundary layer. The boundary layer radar profilers, especially when complemented by temperature profiles generated by a Radio-Acoustic Sounding System (RASS), offer a source of relatively untapped data for model evaluation.

Cloud base height measurements from ceilometers are also reasonable PBL depth indicators for non-clear sky conditions. A spatially extensive network for broad application is available through the NOAA Automated Surface Observing System (ASOS). In addition, since 2004, over 400 commercial aircraft have been collecting meteorological variables (temperature, pressure, relative humidity, winds) as a part of the Tropospheric Airborne Meteorological Data Reporting (TAMDAR) system (http://www. airdat.com/technology/tamdar-sensor-network/). While TAMDAR is designed to provide near-real-time data for forecasting, the system provides valuable vertical profile temperature data (and other variables) during ascents and descents. These data can potentially be synthesized to fill temporal and spatial gaps in ground- based profilers.

The Meteorological Assimilation Data Ingest System (MADIS - http://madis.noaa.gov/) is an integrated system incorporating observations from a variety of surface-based, vertical profile, and satellite networks. MADIS ingests data files from NOAA and non- NOAA sources, decodes the data, and then encodes all of the observations into a common format with uniform observation units and time stamps. These surface and aloft meteorological data provide a finer density, higher frequency, observational database for use as inputs to air quality modeling systems that are designed to characterize air chemistry. Air chemistry and meteorological observations are used in combination for diagnosing air quality model behavior.

Maintaining and Advancing Observation Programs

Emerging challenges in air quality management will require pollutant monitoring programs to become both more comprehensive – that is, they will have to fill existing gaps -- and more integrated.

Measurement Gaps of Specific Species or Parameters

Nitrogen Species

Nitrogen chemistry plays an important role in a variety of environmental problems, such as ozone, particulate matter, acidification, eutrophication, and visibility. Unfortunately, an adequate observation base does not exist to determine if ambient nitrogen responses are consistent with measured and predicted changes in NOx emissions resulting from recent regulations. The ability of the existing urban-oriented measurement network to detect ambient NOx changes associated with regional scale emission reductions from power stations is compromised by strong local NOx emissions. Also, NO_2 data from most network NOx monitors is affected by other oxidized nitrogen species. The NCore network will provide a modest advancement by measuring reactive nitrogen (NOy) in over 70 locations and, together with a more stringent NO_2 NAAQS (EPA, 2010), should spur greater coverage and deployment of instruments producing true NO_2 observations. True NO_2 is an important diagnostic species for atmospheric chemistry processes. These data are needed to validate satellite NO_2 observations and develop scaling of column data to boundary layer concentrations, facilitating the use of satellite data for locations without ground monitors.

Further measurements of oxidized nitrogen species, including peroxy acetyl nitrate (PAN) and nitric acid (HNO_3), would assist diagnosis of deposition and ozone production during atmospheric transport. Increased use of biofuels will potentially elevate PAN concentrations, and, along with observations of other carbonyl compounds, PAN will be an important indicator of the air quality impact of new fuels. Additionally, HNO_3 can be a key indicator species for understanding the NOx versus VOC limitation on ozone production in a given area.

Two reduced nitrogen species, gas-phase ammonia (NH_3) and particulate ammonium ion, are important components of nitrogen mass balance and for assessments of visibility, fine particles, and ecosystem deposition. There are few ambient NH_3 measurements, as most monitoring of this species occurs in strong source (agricultural) locations to estimate emissions flux. Also, ammonium ion is analyzed as part of the chemical speciation program, but NH_3 volatilization creates a negative bias in those values.

Routine measurements of nitrate radical (NO_3), the dominant nighttime oxidizer, at one or two representative locations would enable diagnosis of model predictions of overall nitrogen characterization.

Nitrous acid (HONO) is an important precursor of hydroxyl radicals (Stutz et al., 2004, Zhou et al., 2002), which are critical in daytime atmospheric chemistry. HONO also reacts heterogeneously with aerosols. Its sources and chemistry are not well understood, and are likely to be poorly characterized in air quality models. Among these needs, a reasonable priority would be to enhance measurements of true NO_2 and NH_3 in our national networks.

CO and SO_2

The atmospheric lifetime of carbon monoxide (CO) is one to three months, making it a useful tracer for evaluating emissions and physical process approximations in air quality models. Sulfur dioxide (SO_2) is the predominant precursor of sulfate, which is a major contributor to PM, acid precipitation, and regional haze. The available SO_2 and CO measurements are largely urban and often in proximity to major sources, limiting their representativeness for broader areas. In addition, most of the current instruments were designed to capture high concentrations for compliance purposes, and are not designed to measure the lower concentrations typical of rural areas. Column amounts of both CO and SO_2 available from sensors on current NASA research satellites are planned to be sustained on NOAA JPSS. In addition, the hourly CO measurements planned for GEO-CAPE will provide improved sensitivity to near-surface CO. The value of total column CO and SO_2 from satellites would be enhanced by more spatially rich surface observations.

Mercury

Mercury has significant impacts on ecosystems and human health, but its chemistry in the atmosphere remains unclear. MercNet is a planning effort organized through NADP to standardize and network multi-media mercury measurements. This network will combine existing and new monitors to the extent possible. Speciated mercury measurements are important for model evaluation and tracking of emission reduction strategies, although the existing technology presents challenges in transitioning from research grade instruments to routine operations.

Volatile Organic Compounds

Biogenically generated VOCs (isoprene, terpenes, sesquiterpenes) contribute significantly to the formation of ozone and secondarily formed $PM_{2.5}$. These compounds are not monitored by the urban-based PAMS or toxics networks, which are the primary sources of VOC data in the United States. The absence of VOC data in most moderately sized cities raises concerns regarding the overall representativeness of a network that is based primarily on the severity of ozone problems in the early 1990s. More troubling is the lack of rural VOC data, especially formaldehyde, a designated hazardous air pollutant (HAP) that is also used as a proxy for biogenic emissions and a useful diagnostic for model evaluations. Formaldehyde levels could also indicate alterations in atmospheric chemistry resulting from a future transition to alternative transportation fuels (e.g., alcohols or natural gas). The availability of more spatially rich surface observations would complement the value of total column formaldehyde measurements from satellites.

Organic PM Composition

Planned programs to reduce emissions of inorganic precursors from mobile sources and power plants will increase the organic carbon fraction of the total aerosol budget. This effect is accentuated by large, uncontrollable organic emissions from biogenic and biomass burning sources. Chemical speciation networks provide an aggregated total organic carbon estimate, since it is not practical to resolve the full molecular spectrum of organic aerosols. Nevertheless, key molecular markers would assist source apportionment and distinguishing primary and secondary aerosol. Additional monitoring sites with representative mixes of aerosols and aerosol sources will be needed to supplement or replace current routine speciation analyses.

Aerosol Physical Properties

Improved particle property measurements have been motivated by interest in near roadway and ultrafine particle exposures, particle nucleation processes, and tracking changes in aerosol size distributions associated with alternative transportation fuels (Wahlin et al., 2001). Recent advances in instrumentation produce relatively reliable and low cost estimates of particle number and surface area. These offer potential for application to routine network operations. This will necessitate additional permanent sites that capture long term changes in particle size characteristics. Particle size measurements could be incorporated into a more focused effort to characterize the range of particle and gaseous attributes associated with the near roadway environments.

Spatial Gaps

Integrated assessments necessarily deal with the behavior of pollutants over multiple spatial scales. This is because physicochemical processes occur on overlapping scales of time and distance. Matching actual pollutant exposure to individual humans requires monitoring at a finer spatial scale than provided by current networks. This is a key link in the source-to-outcome accountability chain. Primary emitted pollutants are subject to very dramatic gradients in the near-source region, including most of the 187 designated HAPs as well as a significant fraction of PM. These gradients often coincide with high population density. Characterization of regional- to urban-scale pollutant gradients provides insight on the relative contribution of regional and local sources to local pollutant levels.

Juxtaposed with the need for finer-scale monitoring is an emerging understanding of long range transport and a gradual rise in background pollutant levels. Rising background levels have resulted from increases in world-wide anthropogenic pollutant emissions. The relative significance of these background contributions to U.S. air quality is increasing, given the progress in North American pollution abatement relative to the increased atmospheric loading from expanding economies in Asia. As a result, monitoring pollutant flow across North American borders, as well as global tracking of "background" levels, has become increasingly relevant to North American multi- pollutant air quality management and accountability.

Spatial gaps in U.S. observation networks include:

Near Source - Fine Scale Characterization -- Ambient monitoring networks typically provide primary data to support a broad range of exposure, epidemiological, and risk assessment studies relating pollutant exposures to health outcomes. Epidemiological studies traditionally use air monitoring data from single, centrally- located urban stations as a surrogate for human exposure. However, urban environments in North America often have large and variable pollutant gradients raising uncertainties in outdoor and personal exposures. Neither of these issues is addressed effectively in current monitoring programs.

Internal Rural Coverage -- Three national networks form the backbone of rural air quality measurements: IMPROVE, CASTNET, and the NADP (Section 2). Although these networks were designed for specific objectives, they have also proven very useful for general air quality model evaluation and transport assessments. However, major spatial gaps exist in monitoring of surface-based ozone and key source indicator and precursor species (CO, SO_2, VOCs, speciated aerosol, NOx, and NOy) throughout the mid-section of the Nation (Figure 1 in Section 2). EPA's primary NAAQS, which are set to protect public health, have led to an urban focus in monitoring these species. Rising background pollutant levels (Cooper et al., 2010) have resulted from increases in world-wide anthropogenic pollutant emissions. This is particularly relevant to western mountainous regions, which are confounded by a mix of long distance pollutant transport, stratospheric ozone intrusion, increasing localized emissions from energy exploration and extraction operations, and population growth.

Sentinel Stations to Link Transport Regimes -- The addition of two or three remote stations on the east and west coasts of North America would support trans-oceanic transport assessments, global and regional air quality model evaluation, boundary conditions for models, and insight on background air quality trends. Sentinel sites need to be supported by a stable resource base, since their most significant benefit is often derived from analyzing long-term trends. Coincident measurements are needed of O_3 and aerosol components (nitrate, sulfate, organic and elemental carbon, trace metals), precursors of O_3

and aerosol (total reactive nitrogen, PAN, VOCs, and SO_2), and atmospheric tracers (such as CO, CO_2, and mercury). Because transported pollution is mostly aloft, sentinel stations are especially effective when they are located at altitude and/or include vertical profile measurements.

Vertical Profiles of Key Atmospheric Species -- Vertical profiling of boundary layer and free troposphere air chemistry in North America is limited to lidar networks, specialized field campaigns, and a small number of ozone sonde releases. The CALIOP instrument provides lidar profiles from space, producing several narrow "curtains" per day over the United States. More routine boundary layer profiling of meteorology and air chemistry would provide valuable support for model evaluation and emerging efforts to integrate models and observations.

Surface-based networks typically monitor the lower ten meters of the atmosphere, where most of the air we breathe is located. There are usually significant differences between this air and the PBL, which models attempt to characterize as a homogenous system. Coverage of these vertical transitions is difficult.

The rawinsonde network lacks adequate temporal resolution to adequately track the diurnal patterns of PBL heights, while NPN radar profiling does not provide sufficient vertical resolution for PBL characterization. Moreover, radar profilers have inadequate spatial coverage and lack consensus methodology to synthesize raw data into derived PBL heights conducive to model evaluation. Similar gaps were noted by a recent National Academies report on climate and weather observations (NRC, 2009b), which listed PBL height, air quality measurements above the PBL, and vertical profiles of humidity among the "highest priority observations needed to address current inadequacies."

Since satellite total column data do not simply correspond to surface conditions, routine vertical profiles of key species such as ozone, NO_2, CO, SO_2, and aerosols are needed to establish the relationship between surface-based point and satellite observations. This would increase the value of each system, and help satellite data fill gaps in sparsely monitored areas. Potential investments in vertical profiling programs to help leverage satellite data include:

- Expansion of NOAA's ozone sonde program to provide added spatial coverage in the continental United States, along with the addition of key trace gas measurements.

- A sustained U.S.-based aircraft campaign (national and international flights), similar to the European IAGOS (formerly MOZAIC and CARIBE) effort, to produce routine vertical profiles of key trace gases and aerosols.
- Deployment of fixed-site lidars (aerosol and/or ozone) at key locations throughout North America to provide continuous profiles of aerosol back- scattered light and ozone which can provide a direct link between ground-based, in-situ samplers and column densities from satellite instruments. NASA and NOAA are developing a small number of fixed-site and field-deployable ozone lidars through the Tropospheric Ozone Lidar Network (TOLNet) to assess the potential of such systems. Such a network could build on and complement the existing NDACC lidars and semi-routine aircraft-based measurements by NOAA.

Temporal Gaps

Temporal gaps in measurements include challenges in harmonizing continuous and gravimetric PM mass monitoring, and the lack of continuous or daily speciated particulate observations. As noted elsewhere, measurements from most satellites and sondes are only available once or twice per day, limiting their usefulness.

However, TEMPO in geostationary orbit will begin providing hourly satellite observations for several species when it launches later this decade.

The demand for higher temporal resolution for particulate matter observations has increased as a result of findings regarding human health response (Peters et al., 2001), and with our developing understanding of multi-scale atmospheric processes. With the exception of PM mass, particle properties are not monitored continuously on most networks. Consequently, surface based observations often miss parts of episodic events associated with fires and dust storms, in addition to limiting the number of pairings relating air quality to health outcome records.

While North American networks have deployed over 500 routinely operating continuous $PM_{2.5}$ mass samplers, harmonization of continuous and gravimetric (i.e., Federal reference) methods for PM mass remains a challenge. Measurement artifacts associated with the filter-based, gravimetric techniques creates significant ambiguity in the PM data, such as loss of mass of semi-volatile constituents. Harmonization could make continuous data more comparable with gravimetric data, allowing characterization of particle concentration distributions across large areas, but this could detract from efforts to produce "true" atmospheric aerosol measurements. Correlation techniques could avoid this problem. Eventually, harmonization of these measurements could address both temporal and spatial gaps in $PM_{2.5}$ monitoring.

Routine PM chemical speciation networks acquire 24-hour averaged samples, collected every third or sixth day. This sampling design is adequate for supporting the annual $PM_{2.5}$ standard and the U.S. regional haze program. However, the timing schedule limits the investigation of PM associations with adverse health effects, evaluation of emissions, development of air quality models, and application of source attribution techniques. Continuous PM speciation technology has been incorporated in the Supersites program, and light absorbing aethalometers (an indicator for elemental carbon) are included in the U.S. air toxics NATTS. Ten to twenty continuous sulfate and organic carbon analyzers are located in a mix of SEARCH and state or local agency platforms.

Satellite Observations

As surveyed in Section 2, satellites provide nominally global, daily monitoring of a number of species important for air quality and climate analyses. Recognizing the limitations noted previously, satellites can complement other monitoring systems. Both satellite and ambient observation systems become more valuable when they are integrated. Satellites observe parameters relevant to air quality where little or no routine in-situ monitoring occurs, such as over oceans and rural areas. Low Earth orbit missions provide near-global coverage, with limited temporal resolution and coverage.

Geostationary missions offer much better temporal coverage of a defined region and will begin acquiring observations over much of the Northern Hemisphere by the end of the decade. Although a tremendous advance will occur when TEMPO begins providing many of the measurements planned for GEO-CAPE in approximately 2018, a remaining challenge is to obtain the key companion observations, including CO and CH_4.

Current and future air quality satellite missions provide a stream of parameters directly relevant to air quality information for North America, and should strongly influence the design of our routine monitoring programs. These satellite data will improve near-term air quality characterizations and offer the potential to enhance air quality assessments. However, considerably greater value is realized from satellite observations when they are integrated with complementary ground-based point and vertical profile observations. This begins with validation efforts, when the algorithms used to produce column densities from raw satellite readings are refined and calibrated. Validation depends on independent measurements underneath the satellite. Several of the surface-based recommendations discussed above for trace gas measurements, such as formaldehyde, NO_2, CO, and SO_2, would be very helpful in validation. Intensive field studies have been useful to provide vertical information for satellite validation, but they have limited temporal and spatial coverage.

When validated, satellite data become useful for addressing spatial and/or temporal gaps in ambient monitoring through integration with existing monitoring networks. The integration of satellite information enhances both regional and global scale air quality characterizations, addressing important criteria pollutants such as O_3 and $PM_{2.5}$, and selected air toxics such as formaldehyde.

In addition to the technical challenges and limitations of using satellite data discussed above, it remains organizationally challenging for agencies, such as EPA, to effectively support planning and development of satellite missions.

Integration Opportunities

The assessment of any particular air quality issue, or related set of issues, is best served by the use of many types of observations and models. Likewise, these individual observations and data streams can serve multiple objectives, for disparate organizations and user communities. The integration of observations of different types, from different organizations, is both a major opportunity and a significant challenge.

Conceptually straightforward integration opportunities include:

- Enhancing the horizontal and vertical characterization of key species:
    - horizontally by combining urban- and rural-based networks (e.g., urban-based speciation networks with the rural-based IMPROVE program; SLAMs (urban) and CASTNET (rural) O_3 stations);

  - vertically through the atmospheric column by blending surface measurements, vertically-resolved observations from ground-based and aircraft platforms, and satellite data;
- Combining precipitation and dry observation networks to develop deposition fields, as performed currently through the CASTNET and NADP programs;
- Collocating atmospheric deposition observations with soil and surface water measurement campaigns and select long-term ecological monitoring sites, such as NEON and the Long Term Ecological Research (LTER – www.lternet.edu) Network;
- Collocating a variety of different species measurements to yield multiple pollutant characterizations within a consistent spatial frame;
- Matching ambient measurement fields to human activity patterns to estimate exposures, and matching ambient measurements to emissions fields, via inverse modeling, to refine emission estimates;
- Using air quality models in combination with observations to address spatial and temporal gaps associated with limited observations.

Some of these examples are straightforward, like-to-like combinations of two networks with different but overlapping domains. Others, such as blending of satellite, surface, and aircraft data, or combinations of models with observations, are technically and scientifically challenging. Finally, organizational barriers exist for all of these recommendations, particularly for those that require moving monitors or modifying methods.

Observations and Models to Improve Environmental Characterization

Increasingly, models and observations are being used together, in a variety of ways, due to advances in computational efficiencies and in response to the complexities discussed in Section 1. Air quality models have typically been used in prognostic applications that address hypothetical questions regarding the effects of management programs and rules on future emissions and air quality changes – a function outside the scope of this observations report. More recently, air quality models have been used for forecasting (next day) air pollution, and providing more spatial texture beyond central site monitors to drive human exposure models. Observations alone lack adequate resolution (space, time, and composition) to support integrated assessments.

The integration of chemical observations and transport models is evolving, sharing common attributes with weather characterization and forecasting but less technically mature. Observation and model integration efforts range from using measurements to evaluate model performance, to four dimensional data assimilation (FDDA), as used in meteorological models.

Variations and intermediate levels of integration exist. Areas of observation-model linkages include:

- Model results to guide monitoring site design for placement in areas of expected high concentrations, steep concentration gradients, and important transport corridors;
- Observations supplying direct inputs for initial and/or boundary conditions in models;
- Observations to indirectly improve model inputs through inverse modeling of emissions;
- Observations to evaluate model performance, diagnose model behavior, and constrain model adjustments;
- Observations combined or "fused" with model estimates to add spatial, temporal, and compositional texture to air quality gradients; and,
- Dynamic assimilation of observations to nudge model estimates, analogous to FDDA in meteorological systems.

The linkages between observations and models are emphasized here as a means to influence a shift in monitoring design to explicitly recognize these relationships.

Information Technology to Facilitate Data Access, Integration, and Use

Integrated air quality management requires technology to facilitate discovery, access, handling, archiving, and harmonization of numerous disparate information sources. Accessing and handling observations from single networks or databases remain a challenge, despite large investments. Some applications and tools to access and integrate multiple data systems have eased the integration of disparate data from multiple programs, but by and large differing formats, standards, and gaps in metadata significantly hamper integration of different types of data and models. This prevents analysts from realizing the full value of air quality data.

The EPA system for accessing air quality observations was designed primarily as a repository for data, and covers only a part of U.S. observational archives. Several other organizations have recently built publicly accessible, user-friendly, air quality data reduction, integration, analysis, and visualization systems. These include VIEWS, the Visualization Information Exchange Web System developed by the Regional Planning Organizations (RPOs; http://views.cira.colostate.edu/web/) in support of visibility assessments, the Health Effects Institute's air quality database (Section 2), and ADAM (Airborne Data for Assessing Models; NARSTO (http://www.narsto.org/) also has constructed an accessible database for intensive field campaigns of historical interest.

GEOSS, the Global Earth Observation System of Systems (Appendix A), is a current effort to build a framework to enable national governments to make Earth science data more accessible and usable for decision support.

GEOSS is designed to make data easier to find and access, as well as to support a service-oriented, interoperable, systems approach. This is in contrast to the end-to-end systems typically built to process, handle, and visualize air quality and satellite data. The GEOSS approach is designed to produce more cost-effective, nimble, and usable tools to allow analysts to integrate different types of monitoring data, models, and emissions inventories, etc. This approach has been demonstrated and piloted by several Federally supported projects, including:

- DataFed (http://datafedwiki.wustl.edu/index.php/DataFed_Wiki), which mediates between autonomous, distributed, air quality data providers and users. This facilitates access to and flow of atmospheric data from provider to users, supporting the development of user-driven data processing value chains and improving inputs to comprehensive integrated environmental assessments.
- GEOSS Architecture Implementation Pilot (AIP) Air Quality Workgroup, which developed the GEOSS infrastructure for air quality data and built "Air Quality Community Infrastructure" to serve as an interface between air quality analysts and GEOSS. https://sites.google.com/site/geosspilot2/air quality-and-health-working-group
- Cyberinfrastructure for Air Quality Management (CyAir), which contributes to the planning, development, maintenance, and

> coordination of systems to help the air quality community better utilize air quality information. The cyberinfrastructure is envisioned as a service-oriented, open-source, web-based network of air quality and pollutant emissions data providers and repositories, supported by existing and new data analysis tools for use by the air quality management and research communities. http://cyair.net

These emerging integrated systems will help address technology needs for comprehensive assessments, but will require substantial, sustained investment and engagement from supporting and user communities.

Barriers to Progress

Any approach to addressing the emerging air quality and assessment issues must recognize the resource, technological, and institutional constraints that impede the progress of air quality monitoring programs.

Sustaining Infrastructure

As documented in many fields of endeavor, it is challenging for organizations to maintain infrastructure. Monitoring systems are no different. Users are often in different organizations than providers, and take the data for granted. Indeed, the provision of seamless, automated access to data lets users work with minimal awareness of who produced the data.

Many measurement networks struggle with outdated technology, old equipment, and aging workforces. Downstream assessments are more visible than the monitoring systems that support them, the latter being sustained by a "trickling down" of resources.

Organizational Priorities

Organizations often lack the resource flexibility to support medium or low priority measurements. For example, EPA relies on Federal Reference Methods (FRM) and Federal Equivalent Methods (FEM) to assess compliance with air quality standards. Under constrained budgets, new and improved measurements of high scientific value are generally difficult to fund. A recent example is the continued acceptance of existing NOx instruments with known biases, despite development of a new NO_2 standard.

Transitioning Research and Technology Development to Operations

Measurement programs supported by research organizations are particularly vulnerable to loss of funding, compromising long-term records and other applications. In the case of satellites, "research" sensors/platforms with finite lifetimes are typically not replaced. Although analysis of long-term air quality patterns is a research interest, research organizations typically focus on methods development and physicochemical process characterization. The expectation is that routine measurement programs will transition to operational organizations. For example, NASA satellite missions have defined operational time spans, yet the transition to longer term operational status through partner agencies is generally not planned in advance. Successful transitions have included the LANDSAT mission partnership between USGS and NASA, as well as EPA's management of CASTNET, which was transitioned from EPA's research office to EPA's air office in the late 1990s.

The original NCore monitoring strategy for proposed Level 1 sites had been to form partnerships between universities and state and local agencies to test emerging instrumentation, and jointly share in the transition of research grade equipment to operations. However, Level 1 sites have not been deployed, even though more than 70 Level 2 multiple pollutant measurement sites relying on routine instruments have been deployed. As a result, there has been inadequate incorporation of continuously operated speciated particulate matter, mercury, and inorganic nitrogen species (reduced and oxidized forms) measurements.

Challenges of Long-Term Support for Future Satellite Missions

A particular research-to-operations challenge is the difficulty that recipient agencies face in supporting the development and funding of future satellite missions. Satellite instruments are planned, developed, and funded years before the platform is launched, due partly to the high costs associated with satellite missions. This time frame often extends beyond the planning time horizon of a regulatory agency.

EPA does not fund satellite missions, but it would be useful for it and other recipient agencies to have a mechanism to identify how specific instruments/missions will jointly help monitor and manage ambient air quality. Although it is difficult for a recipient agency to commit to use data many years in the future, such commitments could influence mission funding and instrument design decisions. Conversely, satellite instruments have proven essential for air quality analysis and management, and the provision of data to address emerging air quality assessment challenges should remain high in mission agency planning priorities.

Market Incentives

Beyond the need for FRM/FEM instruments, there are few market incentives for instrumentation firms to pursue the engineering and development steps necessary to produce operational grade methods. This financial barrier is linked to the above noted issues regarding agency priorities, transition from research to operations, and communication impediments between regulatory agencies and instrument developers over anticipated future needs.

Observation technology is typically developed by individual research groups for specific applications associated with a laboratory or field campaign objective. This technology can then be passed on to other users. For example, NASA develops satellite observation platforms in space and on aircraft typically for single-use (or short-term), promising technologies. These are then transferred to NOAA, and the satellite sensors made operational by NESDIS. There is no similar development path for technologies to inform surface air chemistry monitoring for use by EPA or state and local air quality managers.

CONCLUSIONS

The United States has a robust and invaluable network of air quality observation systems, and recent improvements in technology are providing unprecedented opportunities to enhance current capabilities. However, there also exist substantial opportunities for improvement in the U.S. air quality observation system as currently implemented.

Establish Standing Multi-Agency Observations Working Group

As a first step to taking advantage of these opportunities, the NSTC should consider chartering an interagency working group on air quality observations under the AQRS of the CENRS. This group would maintain close coordination with national Civil Earth Observations planning and the U.S. Group on Earth Observations (USGEO) activities, including consideration of existing partnership models (e.g., NADP, IMPROVE). The objectives of this working group could include:

- Provide a forum to facilitate cooperation and collaboration among the Federal agencies with air quality observation programs. Air quality measurements are important to so many users that a broader view of the health, relevancy, and evolution of observation programs should complement the existing single-organization focus on discrete network elements.
- Provide an interface between the various user communities (e.g., air quality managers, health scientists, air quality forecasters, etc.) and those involved in making air quality observations to ensure the benefits are maximized for both communities.
- Extend the analysis presented in this report and develop specific recommendations for improvements to the Nation's air quality observing system and track progress towards the goals established. Specifically, the Working Group on Air Quality Observations would:
  - assess the adequacy of current networks and measurement technologies, including maintenance shortfalls and modernization needs
  - identify important measurement gaps
  - identify important information gaps and opportunities for advancing technology and sharing and utilization of observation programs.
- Coordinate the development of multi-agency initiatives to address deficiencies that have been identified and enhance and extend air quality observations in the U.S.

Address Current Observation Gaps

Numerous important measurements that are missing or in short supply were described in Section 3. This analysis will need to be revisited as monitoring systems and our understanding of the atmosphere evolve, and it will be appropriate for the working group to add their perspective to the analysis presented there.

While requests for added observations have been raised periodically, this renewed effort is intended to (a) increase the overall value-to-cost ratio incurred collectively through a system of measurement programs and (b) improve the comprehensive effectiveness of measurement programs where past requests have focused on specific topics without recognition of the broader opportunities for leverage and cooperation. Suggested steps include:

1. initiate monitoring of reactive gas and particulate nitrogen compounds, which are precursors of ozone and particulate matter, contributors to acid deposition, and nutrients in ecosystems,
2. collocate instrumentation at core monitoring sites to facilitate inter-comparison with satellite observations,
3. target monitoring in rural/remote areas to measure regional backgrounds and contributions from long-range transport of pollutants,
4. establish more robust air toxics monitoring near major industrial facilities to help investigate whether air toxics emissions are associated with human health effects in nearby communities,
5. target intensive field studies designed to elucidate critical processes that determine atmospheric concentrations of ozone and particulate matter and other air pollutants, and
6. establish routine monitoring of vertically resolved observations of ozone, fine particulate matter (including its composition), and their precursors, to evaluate and improve air quality modeling.

Introduction to Appendices for Existing Air Quality Monitoring Programs

Observation programs supporting air quality and related assessments include routine regulatory networks, deposition networks, intensive field studies, remote sensing systems, sondes, aircraft campaigns, satellites, and focused fixed-site special purpose networks. Appendices A – I provide information on a wide variety of these air monitoring networks. Major networks that are currently operating are emphasized; in some cases, reference to other networks that have been discontinued or were intended only for a specific operating period is also provided. The focus is on networks located in the United States, but attention is also given to other North American, European, and international efforts that contribute to U.S. assessments.

Information on monitoring networks represents the recent status of these networks (in terms of monitor number, placement, and measurement parameters). The actual status will likely have changed as of the publication date of this report for at least some of these networks as they expand or contract, consistent with current needs. For latest information on the status of many air quality monitoring networks, readers should consult AirData at http://www.epa.gov/airdata/ad_maps.html. Alternately, website links (which also periodically change) in these appendices should be consulted for the latest information on networks.

This information is the product of extensive Internet searches and information provided by knowledgeable representatives from the agencies responsible for the networks. In most cases, the information provided has been taken directly from the referenced Internet site, especially for supplemental information for the non-routine special intensive studies. Attribution of this information should be to those Internet websites.

APPENDIX A. INTEGRATED OBSERVATIONAL STRATEGIES

Intensive field campaigns and observations from space provide valuable measurements not captured in routine, ground-based observation networks and can have direct impacts on the air quality management process. For example, results from the Texas Air Quality Study were incorporated into the ozone State Implementation Plan for the Houston area within two years. Similarly, satellite observations have played important roles in improving the quality of fire-based emissions estimates in EPA's systems for driving air quality models. And in the absence of an adequate surface-based network, satellite observations have been used to demonstrate the progress of major national programs to reduce emissions of nitrogen oxides.

Integrating these various types of observations with one another and with models to produce the best multi-dimensional characterization of air quality is a significant scientific and information-technology challenge, even in a research setting. Integration for decision-support purposes, near-real-time forecasting applications, and other operational purposes is even more challenging.

Over the last several years, a number of observational strategies and umbrella organizations have formed that convey and promote integration across disciplines, observational modalities, and/or organizations. Some of these efforts are focused on air quality or atmospheric chemistry, while some are far broader. These strategies and organizations include:

USGEO and GEO – The U.S. Group on Earth Observations (USGEO) exists as a subcommittee of the NSTC Committee on Environment, Natural Resources, and Sustainability (CENRS). USGEO's mandate is to: (1) coordinate, plan, and assess Federal Earth observation activities in cooperation with domestic stakeholders; (2) foster improved Earth system data management and interoperability throughout the Federal Government; and (3) engage international stakeholders by formulating the U.S. position for, and coordinating U.S. participation in, the intergovernmental GEO based in Geneva, Switzerland. GEO, a voluntary partnership of national governments and international organizations, is developing the Global Earth Observation System of Systems (GEOSS). GEOSS provides a voluntary, multidisciplinary framework to make all types of Earth observations more discoverable, accessible, and useable for decision support. In October 2010, Congress charged OSTP with a new "mechanism *to ensure greater coordination of the research, operations, and activities relating to civilian Earth observation*" [emphasis added] that would also produce and routinely update the strategic plan for Earth observations.[1] In response, OSTP developed and released a U.S. National Strategy in April 2013 and concurrently performed an internal assessment of 362 Federal Earth observing systems.

Summary results of this assessment will form the foundation for a new U.S. National Plan for Civil Earth Observations, to be published in 2014. This air quality observations report, along with other reports of the NSTC and the National Academies, provides important technical input to the development of the National Plan.

IGAC/AC&C – The International Global Atmospheric Chemistry (IGAC: http://www.igacproject.org/) program was created in the late 1980s to address growing international concern about atmospheric changes. IGAC is jointly sponsored by the Commission on Atmospheric Chemistry and Global

Pollution (CACGP) of the International Association of Meteorology and Atmospheric Sciences (IAMAS) and the International Geosphere-Biosphere Programme (IGBP). IGAC has initiated or coordinated much of the research over the last decade focusing on chemical composition, transformations, and transport in the troposphere. Together with the Stratospheric Processes And their Role in Climate (SPARC) project of the World Climate Research Programme (WCRP), IGAC has started the Atmospheric Chemistry and Climate (AC&C; http://www.igacproject.org/node/8/) initiative, which examines the interplay between chemistry, chemically-active species, and climate change.

IGACO – Integrated Global Atmospheric Chemistry Observations (IGACO; http://www.igaco-O_3.fi) is a strategy for bringing together ground-based aircraft and satellite observations of 13 chemical species in the atmosphere. IGACO will be implemented as a strategic element of the Global Atmospheric Watch (GAW) program of the World Meteorological Organization (WMO). IGACO will be organized around four focus areas, one of which is air quality / long-range transport. IGACO provides specific recommendations on measurement parameters, and facilitates integration across satellite- and ground-based stations. Although IGACO is focused on large, global-scale characterizations, the strategy provides useful guidance that should be considered in any air-based observation program design. Several of the core IGACO measurement parameters (O_3, CO, NO_2, and CO_2) are important regional- and urban-scale air quality indicators.

MACC – Monitoring Atmospheric Composition and Climate (MACC; http://www.gmes.info/pages-principales/projects/atmosphere-projects/macc/) is a recently initiated collaborative effort, funded by the European Commission, to monitor the global distribution and long-range transport of long-lived greenhouse gases, aerosols, and reactive pollutants that degrade air quality. MACC's product lines include data records on atmospheric composition for recent years, and current data for monitoring present conditions and forecasting the distribution of key constituents a few days ahead. (MACC is a continuation of the Global and regional Earth-system (Atmosphere) Monitoring using Satellite and in-situ data (GEMS) and PROtocol MOniToring for the GMES Service Element: Atmosphere (PROMOTE) programs under Global Monitoring for Environment and Security (GMES), see above website for details.)

NARSTO – NARSTO (formerly the North American Research Strategy for Tropospheric Ozone; http://www.narsto.org/) is a public-private partnership of government agencies, industry, and academic institutions. The partnership sponsors a variety of workshops and assessments addressing current air quality research interests. NARSTO has focused on the atmospheric sciences, with assessments addressing ozone and particulate matter air pollution, emissions inventories, and, more recently, multiple-pollutant air quality management. These assessments generally complement preceding NAS studies addressing air pollution management. The NARSTO archive stores data from a variety of intensive field campaigns.

NAAMS – The National Ambient Air Monitoring Strategy (NAAMS; http://www.epa.gov/ttnamti1/monitor.html) was developed jointly by the EPA and numerous state and local agencies. Developed in the early 2000s, NAAMS (Scheffe et al., 2009) was intended to make the design of U.S. regulatory-based networks more efficient in supporting the development of air quality standards and emission control strategies. The multi-pollutant National Core network (NCore, see Section 2) emerged from the NAAMS process.

AQAST – Air Quality Applied Sciences Team (AQAST; http://acmg.seas.harvard.edu/aqast/index.html) is a NASA team of atmospheric scientists working in partnership with U.S. air quality managers from local to national levels. The goal is to exploit the power of Earth science tools and data sets, available from NASA and other agencies, to address multi-faceted air quality problems. A wide range of projects using satellite data, suborbital data, and models are conducted through pooled expertise under the program.

APPENDIX B. EVOLUTION OF UNITED STATES AIR MONITORING NETWORKS

The 1970 Clean Air Act (CAA) established a framework for the original National Ambient Air Quality Standards (NAAQS), and drove the design and implementation of the NAMS and SLAMS networks in the late 1970s. These networks were intended primarily to establish non-attainment areas with respect to the NAAQS, which include ozone, sulfur dioxide, nitrogen dioxide, carbon dioxide, lead (Pb), and particulate matter (PM). The NAMS/SLAMS networks have evolved over time (Figure A.1) as a result of cyclical NAAQS review and promulgation efforts, leading to changes in measurement requirements related to averaging times, locations, and the various size cuts associated with PM.

Relatively wide geographical distribution and persistence of ozone and $PM_{2.5}$ NAAQS dominating the national exceedances (Figure A.2) have led to these pollutants monitoring landscape.

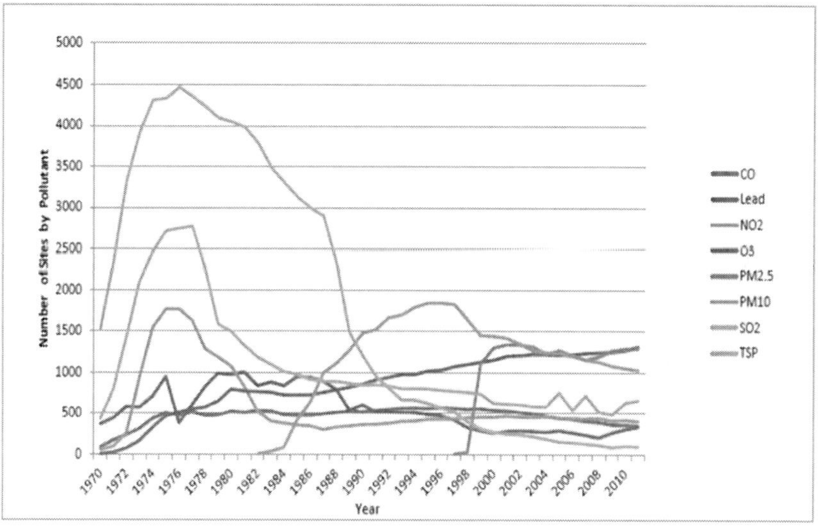

Figure B.1. Evolution of U.S. air network growth.

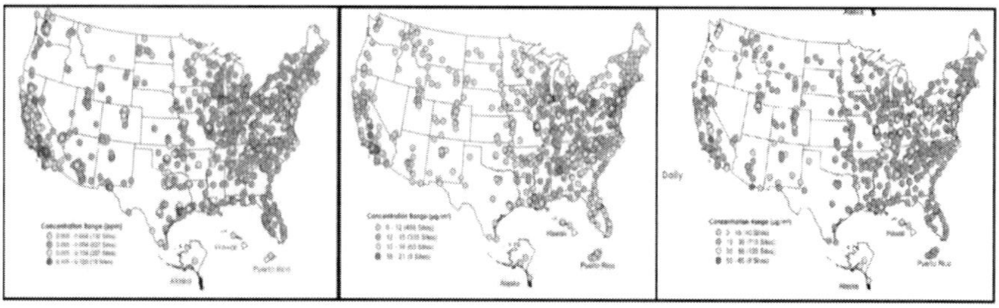

Figure B.2. 2006 air quality summaries for ozone, annual average $PM_{2.5}$, and daily $PM_{2.5}$.

Figure B.3. Regional chemical composition of $PM_{2.5}$ aerosols -- composition of $PM_{2.5}$ for 15 selected urban areas in the United States. Annual average $PM_{2.5}$ concentrations (ug/m_3) are presented, the circle size representing the magnitude of $PM_{2.5}$ mass.

Yellow and red sites indicate values exceeding NAAQS levels (source, EPA). Two important ambient air networks focused on environmental welfare effects were established in the mid -1980's. The Interagency Agency Monitoring of Protected Visual Environments (IMPROVE) network consists of 212 sites (170 current and 42 discontinued) representing 156 visibility-protected federal areas (national parks, wilderness areas, and wildlife refuges). IMPROVE is used primarily to assess visibility impairment, but has provided a reliable long term record of PM mass and major speciation components. It served as a model for the later deployment of EPA's CSN network (Figure 2 of the full report), which has provided an urban complement to characterize aerosol composition (Figure A.3).

The Clean Air Status and Trends Network (CASTNET) was established in the early 1990s to track changes in dry deposition of major inorganic ions and gaseous precursors associated with the CAA Title 4 reductions in sulfur and nitrogen. Title 4 was designed to address surface water acidification in eastern North America. Complementing ongoing precipitation measurements from the National Atmospheric Deposition Program (NADP), CASTNET has provided a valuable source of model evaluation data for many of the large regional scale applications since the 1990's.

Deployment of the Photochemical Assessment and Measurements Stations (PAMS) and the $PM_{2.5}$ networks from the early 1990's through 2002 markedly enhanced the spatial, temporal, and compositional measurement of gases and aerosols. PAMS partially supports user needs beyond NAAQS compliance, such as public reporting and forecasting of adverse air quality, implementation efforts including air quality model evaluation, and source apportionment and pattern (spatial and temporal) analysis of precursor species.

State and local air agencies have measured a variety of metallic and gaseous hazardous air pollutants (HAPs) at over 200 locations since the 1980's. Typically, broad access to and use of data was compromised by a lack of centralized databases and multiple sampling and laboratory protocols leading to data uncertainty. In response to this gap in accessible and centralized HAPs observations, a modest 27 site National Air Toxics Trends (NATTS) network was initiated in 2001. Current NATTS species include: acrolein, perchloroethylene, benzene, carbon tetrachloride, chloroform, trichloroethylene, 1,3- butadiene, 1,2-dichloropropane, dichloromethane, tetrachloroethylene, vinyl chloride, formaldehyde, acetaldehyde, nickel compounds, arsenic compounds, cadmium compounds, manganese compounds, beryllium, lead, hexavalent chromium, and expected additions of benzo(a)pyrene and naphthalene.

A new multi-pollutant monitoring network, referred to as NCore, was incorporated in the 2006 revisions to the particulate matter standards. When fully implemented, NCore will provide a minimum of 75 Level 2 sites (Figure A.4) in most major urban areas and important transport corridor and background locations. NCore will include a variety of trace gas, aerosol mass and speciation measurements which are intended to support multiple data user needs (e.g., air quality model evaluation, long term epidemiological studies). In addition to establishing a multiple pollutant measurement framework, the NCore sites are intended to provide a backbone of central location sites that can be complemented by additional (existing and new) stations to address more specific spatial resolution requirements. Intensive Level 1 sites, intended to promote transition of new technologies into routine networks, have not been implemented to date.

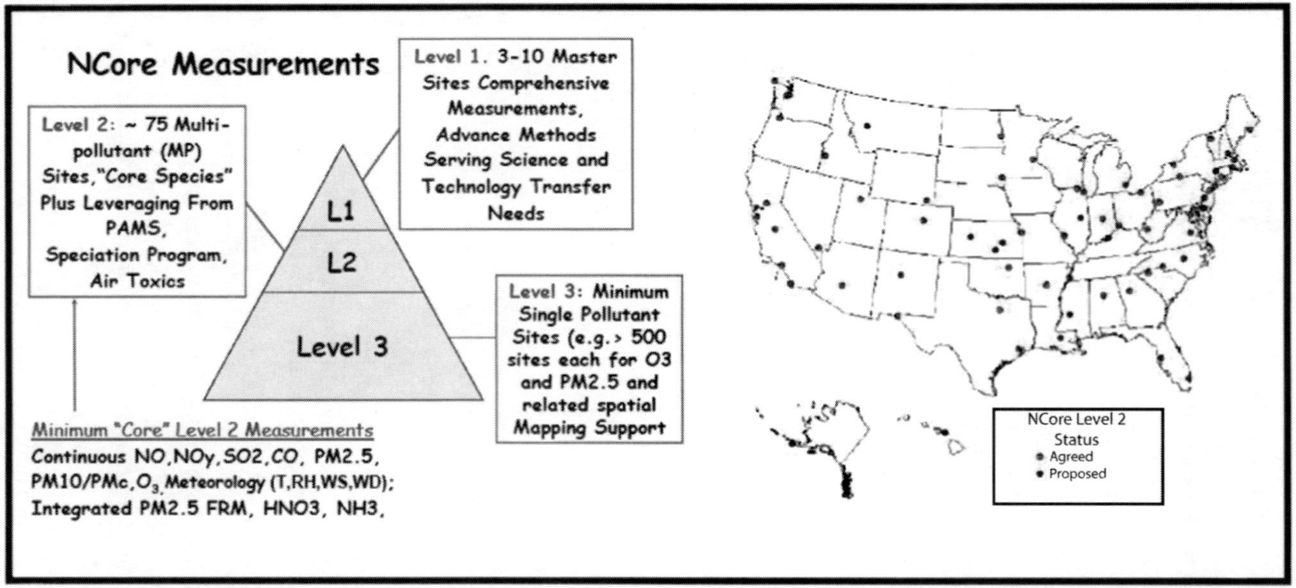

Figure B.4. Original 3-tiered NCore design (left) and proposed site locations (see http://www.epa.gov/ttn/amtic/ncore/networks.html_for updated site locations).

APPENDIX C. MAJOR ROUTINE OPERATING AIR MONITORING NETWORKS[3]

Network	Lead Agency	Number of Sites	Initiated	Measurement Parameters	Location of Information and/or Data
State / Local / Federal Networks					
NCore – National Core Monitoring Network	EPA	80	2008	O3, NO/NO2/NOy, SO2, CO, PM2.5/PM10-2.5, PM2.5 speciation, NH3, HN O3, Surface Meteorology[1]	http://www.epa.gov/ttn/amtic/ ncore/index.html
SLAMS – State and Local Ambient Monitoring Stations	EPA	~1500	1978	O3, NOx/NO2, SO2, PM2.5/PM10, CO, Pb	http://www.epa.gov/air/oaqps/ qa/ monprog.html
CSN – Chemical Speciation Network	EPA	~200	1999	PM2.5, PM2.5 speciation, Major Ions, Metals	http://www.epa.gov/ttnamti1/ specgen.html
PAMS—Photochemical Assessment Monitoring Network	EPA	50+	1994	O3, NOx/NOy, CO, Speciated VOCs, Carbonyls, Surface Meteorology & Upper Air	http://www.epa.gov/ttn/amtic/ pamsmain.html
IMPROVE—Interagency Monitoring of Protected Visual Environments	DOI (NPS)	~160	1988	PM2.5/PM10, Major Ions, Metals, Light Extinction, Scattering Coefficient	http://vista.cira.colostate.edu/ IMPROVE/

(Continued)

Network	Lead Agency	Number of Sites	Initiated	Measurement Parameters	Location of Information and/or Data
CASTNet – Clean Air Status and Trends Network	EPA	80+	1987	O3, SO2, Major Ions, Calculated Dry Deposition, Wet Deposition, Total Deposition for Sulfur/Nitrogen	http://www.epa.gov/castnet/
GPMN—Gaseous Pollutant Monitoring Network	DOI (NPS)	33	1987	O3, NOx/NO/NO2, SO2, CO, Surface Meteorology, (plus enhanced monitoring of CO, NO, NOx, NOy, and SO2 plus canister samples for VOC at three sites)	http://www2.nature.nps.gov/air/Monitoring/network.cfm#data
POMS—Portable Ozone Monitoring Stations	DOI (NPS)	14	2002	O3, surface meteorology, with CASTNet-protocol filter pack (optional) sulfate, nitrate, ammonium, nitric acid, sulfur dioxide	http://www2.nature.nps.gov/air /studies/port O3.cfm

Network	Lead Agency	Number of Sites	Initiated	Measurement Parameters	Location of Information and/or Data
Passive Ozone Sampler Monitoring Program	DOI (NPS)	32	1995 (ended 2005)	O_3 dose (weekly)	http://www2.nature.nps.gov/air /Studies/Passives.cfm
NADP/AMoN—National Atmospheric Deposition Program / Ammonia Monitoring Network	Multi-agency	~50	2010	NH_3	http://nadp.sws.uiuc.edu/AMoN/
NADP/AMNet—National Atmospheric Deposition Program / Atmospheric Mercury Network	Multi-agency	~20	2009	Gaseous elemental mercury (GEM) Gaseous oxidized mercury (GOM) Particulate-bound mercury (PBM2.5) Surface Meteorology	http://nadp.sws.uiuc.edu/amn/
NADP/MDN—National Atmospheric Deposition Program / Mercury Deposition Network	Multi-agency	110+	1996	Mercury from precipitation chemistry	http://nadp.sws.uiuc.edu/mdn/
NADP/NTN—National Atmospheric Deposition Program / National Trends Network	Multi-agency	~250	1978	Major Ions from precipitation chemistry	http://nadp.sws.uiuc.edu/ NTN/
NADP/AIRMoN—National Atmospheric	Multi-agency	7	1992	Major Ions from precipitation chemistry	http://nadp.sws.uiuc.edu/AIRM oN/

(Continued)

Network	Lead Agency	Number of Sites	Initiated	Measurement Parameters	Location of Information and/or Data
Deposition Program / Atmospheric Integrated Research Monitoring Network				Note: some sites began in 1976 as part of the DOE MAP3S program; early data are archived on NADP and ARL servers.	
IADN—Integrated Atmospheric Deposition Network	EPA	15	1990	PAHs, PCBs, and organochlorine compounds are measured in air and precipitation samples	http://www.epa.gov/greatlakes/monitoring/air2/
NAPS—National Air Pollution Surveillance Network	Canada	250+	1969	SO2, CO, O3, NO, NO2, NOx, VOCs, SVOCs, PM10, PM2.5, TSP, metals	http://www.ec.gc.ca/natchem/default.asp?lang=en&n=EE0E2169-1
CAPMoN—Canadian Air and Precipitation Monitoring Network	Canada	33	2002	O3, NO, NO2, NOy, PAN, NH3, PM2.5, PM10 and coarse fraction mass, PM2.5 speciation, major ions for particles and trace gases, precipitation chemistry for Major Ions	http://www.ec.gc.ca/rs- mn/default.asp?lang=En&n=752CE271-1

Network	Lead Agency	Number of Sites	Initiated	Measurement Parameters	Location of Information and/or Data
Mexican Air Quality Network	Mexico	52-62	Late 1960's	O3, NOx, CO, SO2, PM10, TSP,VOC	http://www.ine.gob.mx/dgicur/calaire/indicadores.html
Mexican City Ambient Air Quality Monitoring Network	Mexico	49	Late 1960's	O3, NOx, CO, SO2, PM10, TSP,VOC	http://www.ine.gob.mx/dgicur/calaire/indicadores.html
Air Toxics Monitoring Networks					
NATTS—National Air Toxics Trends Stations	EPA	27	2005	VOCs, Carbonyls, PM10 metals[2], Hg	http://www.epa.gov/ttn/amtic/airt oxpg.html
State/Local Air Toxics Monitoring	EPA	250+	1987	VOCs, Carbonyls, PM10 metals[2], Hg	http://www.epa.gov/ttn/amtic/airt oxpg.html
NDAMN—National Dioxin Air Monitoring Network	EPA	32	1998 - 2005	CDDs, CDFs, dioxin-like PCBs	http://cfpub.epa.gov/ncea/CFM/re cordisplay.cfm?deid=54811
Tribal Monitoring Networks					
Tribal Monitoring[4]	EPA	90+	1995	O3, NOx/NO2, SO2, PM2.5/PM10, CO, Pb	http://www.epa.gov/air/tribal/airp rogs.html#ambmon
Industry / Research Networks					
New Source Permit Monitoring	None	variable	variable	O3, NOx/NO2, SO2, PM2.5/PM10, CO, Pb	Contact specific industrial facilities
HRM Network—Houston Regional Monitoring Network	None	9	1980	O3, NOx, PM2.5/PM10, CO, SO2, Pb, VOCs, Surface Meteorology	http://hrm.radian.com/houston/ho w/index.htm

(Continued)

Network	Lead Agency	Number of Sites	Initiated	Measurement Parameters	Location of Information and/or Data
ARIES / SEARCH—Aerosol Research Inhalation Epidemiology Study / Southeastern Aerosol Research and Characterization Study experiment	None	6	1992	O3, NO/NO2/NOy, SO2, CO, PM2.5/PM10, PM2.5 speciation, Major Ions, NH3, HN O3, scattering coefficient, Surface Meteorology	http://www.atmospheric-research.com/studies/SEARCH /index.html
SOS – SERON—Southern Oxidant Study - Southeastern Regional Oxidant Networks	EPA	~40	1990	O3, NO, NOy, VOCs, CO, Surface Meteorology	http://www.ncsu.edu/sos/pubs/ sos 3/State_of_SOS_3.pdf
National/Global Radiation Networks					
RadNet—formerly Environmental Radiation Ambient Monitoring System (ERAMS)	EPA	200+	1973	Radionuclides and radiation	http://www.epa.gov/enviro/ html/e rams/
SASP -- Surface Air Sampling Program	DHS	41	1963 (no data after 1993)	89Sr, 90Sr, naturally occurring radionuclides, 7Be, 210Pb	http://www.nbl.doe.gov/htm/ EML_ Legacy_Website/databases.htm

Network	Lead Agency	Number of Sites	Initiated	Measurement Parameters	Location of Information and/or Data
NEWNET—Neighborhood Environmental Watch Network	DOE	11	1993	Ionizing gamma radiation, Surface Meteorology	http://newnet.lanl.gov/
Solar Radiation Networks					
UV Index – EPA Sunrise Program[5]	EPA	~50 U.S. cities	2002	Calculated UV radiation index	http://www.epa.gov/sunwise/uvin dex.html
UV Net -- Ultraviolet Monitoring Program	EPA	21	1995/ 2004	Ultraviolet solar radiation (UV-B and UV-A bands), irradiance, ozone, NO2	http://www.epa.gov/uvnet/access. html
NEUBrew (NOAA-EPA Brewer Spectrophotometer UV and Ozone Network)[6]	NOAA	6	2005	Ultraviolet solar radiation (UV-B and UV-A bands), irradiance, ozone, SO_2	http://www.esrl.noaa.gov/gmd/grad/neubrew/
UV-B Monitoring and Research Program	USDA	37	1992	Ultraviolet-B radiation	http://uvb.nrel.colostate.edu/UVB/ index.jsf
SURFRAD – Surface Radiation Budget Network	NOAA	8	1993	solar and infrared radiation, direct and diffuse solar radiation, photosynthetically active radiation, UVB, spectral solar, and meteorological parameters	http://www.srrb.noaa.gov/surfrad/ index.html

(Continued)

Network	Lead Agency	Number of Sites	Initiated	Measurement Parameters	Location of Information and/or Data
AERONET – Aerosol RObotic NETwork	NASA co-located networks	~500	1998	Aerosol spectral optical depths, aerosol size	http://aeronet.gsfc.nasa.gov/ index.
				distributions, and precipitable water	html
MPLNET – Micro-pulse Lidar Network		35	2000	Aerosols and cloud layer heights	http://mplnet.gsfc.nasa.gov/
PRIMENet -- Park Research and Intensive Monitoring of Ecosystems NETwork[6]	DOI (NPS)	14	1997 (ended 2004)	ozone, wet and dry deposition, visibility, surface meteorology, and ultraviolet radiation	http://www.cfc.umt.edu/ primenet/Assets/ Announcements/99PReport.pdf

Footnotes:

1. Surface Meteorology includes wind direction and speed, temperature, precipitation, relative humidity, solar radiation (PAMS only).
2. PM10 metals may include arsenic, beryllium, cadmium, chromium, lead, manganese, nickel, and others.
3. Some networks listed separately may also serve as subcomponents of other larger listed networks; as a result, some double counting of the number of individual monitors is likely.
4. The number of sites indicated for tribal monitoring is actually the number of monitors, rather than sites.
5. Sunrise program estimates UV exposure levels through modeling - does not include measurements.
6. NEUBrew is a subset of Original UV brewer network (UV Net); PRIMENET participated in UV Net program.

APPENDIX D. NATIONAL ROUTINE METEOROLOGICAL MONITORING NETWORKS

Network	Lead Agency	Number of Sites	Initiated	Measurement Parameters	Location of Information and/or Data
ASOS -- Automated Surface Observing System	NOAA	~1000 (supplemented by military weather observation sites)	1992 (replaced routine surface observations collected manually at 260 Weather Service facilities)	Continuous measurements of: Wind Direction and Wind Speed; Visibility; Runway Visual Range; Type, intensity and amount of rain, snow, etc.; Obstructions due to fog, mist, etc.; Cloud Height and Amount; Ambient Temperature; Dew Point Temperature; Pressure; Lightning detection; Automated, manual, and plain language remarks on special weather conditions (depending on level of service); and Additive and automated maintenance data on precipitation amount, max/min temperature, pressure tendency, etc.	http://www.nws.noaa.gov/asos/pdfs/aum-toc.pdf
Cooperative Observer Program	NOAA	~11,400	1890	24-hour maximum and minimum temperatures, Liquid equivalent of precipitation, snowfall, snow depth, and Other special phenomena such as days with thunder, hail, etc.	http://www.nws.noaa.gov/om/coop/

(Continued)

Network	Lead Agency	Number of Sites	Initiated	Measurement Parameters	Location of Information and/or Data
SLAMS – State and Local Ambient Monitoring Stations	EPA	~1500	1978	Wind direction and speed, Temperature, Precipitation, Relative humidity	http://www.epa.gov/air/oaqps/qa/monprog.html
Remote Automated Weather Stations	USDA (USFS)	~2200	~1978	Wind direction and speed, Precipitation, Pressure, Temperature, Relative humidity, Fuel moisture and temperature	http://raws.fam.nwcg.gov/
NOAA Profiler Network (and Cooperative Agency Profilers)	NOAA	35 (plus ~100 CAP sites)	1992	Vertical profiles of wind direction and speed (and vertical profiles of temperature at RASS sites)	http://www.profiler. noaa.gov/npn/
Upper Air Stations (Weather Balloons)	NOAA	~100 in North America, Pacific Islands, and the Caribbean	1937	Measurements of temperature, relative humidity, wind direction and speed, and altitude/height at selected pressure levels.	http://www.ua.nws.noaa.gov/net-info.htm
Forecast Systems Laboratory Aircraft Communications Addressing and Reporting System	NOAA	~4000 commercial aircraft	2001 (routinely available database; discontinued in 2008)	Wind direction, wind speed and temperature reported for various altitudes at which aircraft typically operate	http://www.arm.gov/instruments/acars
National Doppler Radar Sites	NOAA	~160	1990 (national radar network	Base Reflectivity, Composite Reflectivity, One-Hour	http://www.srh.noaa.gov/radar/radinfo/radinfo.html

Network	Lead Agency	Number of Sites	Initiated	Measurement Parameters	Location of Information and/or Data
			originated prior to 1960)	Precipitation, and Storm Total Precipitation	
National Lightning Detection Network	Commercial	100+	1989	Detection of cloud-to-ground lightning flashes at distances up to 400 km	http://www.nwstc.noaa.gov/METEOR/Lightning/detection.h tm
National Environmental Satellite, Data, and Information Service	NOAA	2 GOES satellites 2 POES satellites	1994 (earlier satellite systems replaced)	Vertical profiles of temperature, moisture, and wind; visible and infrared imagery of clouds; water vapor imagery	http://www.goes.noaa.gov/
C-MAN – Buoy and Coastal-Marine Observing Network	NOAA	51	Early 1980s	Pressure, wind direction, wind speed and gust, and air temperature, relative humidity, precipitation, visibility, sea water temperature, water level, and waves	http://www.ndbc.noaa.gov/cman.php

APPENDIX E. EUROPEAN AIR MONITORING NETWORKS

Network	Lead Agency	Number of Sites	Initiated	Measurement Parameters	Location of Information and/or Data
EMEP – Co-operative Programme for Monitoring and Evaluation of the Long-range Transmission of Air Pollutants in Europe (encompasses networks for ~37 European countries and organizations)	UNECE	~200	1977	**Acidifying / Eutrophying Compounds** (precipitation): SO4, N O3, NH4, trace elements, pH, acidity (air): SO2, NO2, HN O3, NH3, PM10, PM2.5, major ions **O3** **Heavy Metals** precipitation, major ions, PM2.5, PM10, Hg, wet deposition **POPs** precipitation, air, deposition **Particulate Matter** PM2.5, PM10, EC, OC, TC, BC **VOC** Hydrocarbons, Carbonyls	http://www.nilu.no/projects/ccc/emepdata.html
EUROTRAC – The European Experiment on the Transport and Transformation of Environmentally Relevant Trace Constituents over Europe	International Executive Committee (European Countries)	N/A	1986	EUROTRAC programs performed analyses utilizing data from existing or specially designed monitoring networks in order to: 1. elucidate the chemistry and transport of ozone and other photo-oxidants in the troposphere, e.g., TOR -- 30 O3 stations and ALPTRAC -- 15 snow monitoring sites 2. identify processes leading to the formation of acidity in the atmosphere, particularly those involving aerosols and clouds.	http://www.helmholtz-muenchen.de/ eurotrac/ index_ what_is.html

Network	Lead Agency	Number of Sites	Initiated	Measurement Parameters	Location of Information and/or Data
				3. understand uptake and release of atmospheric trace substances by the biosphere.	
EUROTRAC-2 – The EUREKA project on the transport and chemical transformation of trace constituents in the troposphere over Europe; second phase. Subprojects: – AEROSOL – BIATEX-2 – CAPMAN – CMD – EXPORT-E2 – GENEMIS – GLOREAM – LOOP – MEPOP – PROCLOUD – SATURN – TOR-2 – TRAP45 – TROPOSAT	International Scientific Secretariat (European Countries and EU)	N/A	1996	EUROTRAC-2 programs performed analyses utilizing data from existing monitoring networks in order to support the further development of abatement strategies within Europe by providing an improved scientific basis for the quantification of source-receptor relationships for photo-oxidants and acidifying substances.	http://www.gsf.de/eurotrac/index_what_is.htm l

APPENDIX F. MONITORING NETWORKS FOR PERSISTENT ORGANIC POLLUTANTS (POPS)

Network	Lead Agency	Number of Sites	Initiated	Measurement Parameters	Location of Information and/or Data
Global Monitoring of Persistent Organic Pollutants (POPs)[1]	UNEP – United Nations Environment Programme	N/A	2003	Activities include developing guidance on sampling and analysis of POPs, QA/QC procedures, data treatment and communication and data assessment. In addition, the program will include an electronic discussion group on POPs monitoring issues where existing programs and laboratories are invited to participate and share their experience on this subject.	http://www.chem.unep.ch/gmn/default.htm
AMAP – Artic Monitoring and Assessment Programme	NOAA (as U.S. representative to the 8 nation Artic Council)	N/A	~1991	Air/aerosol sampling for POPs, heavy metals, radioactivity and acidification parameters; bulk precipitation and snowpack sampling to estimate deposition2	http://www.amap.no/
EMEP -- Co-operative Programme for Monitoring and Evaluation of the Long-Range Transmission of Air Pollutants in Europe	UNECE – United Nations Economic Commission for Europe	17	1991	Benzo(a)pyrene, PCBs, hexachlorobenzene, chlordane, lindane, hexachlorocyclohexane, DDT/DDE in precipitation and gas particles	http://www.chem.unep.ch/gmn/012_emep.htm

Network	Lead Agency	Number of Sites	Initiated	Measurement Parameters	Location of Information and/or Data
GAPS – Global Atmospheric Passive Sampling	UNEP – United Nations Environment Programme	50	2004	12 chemicals including Aldrin, Chlordane, DDT, Deldrin, Endrin, Heptachlor, Hexachlorobenzene, Mirex, PCBs, Dioxins(PCDDs), Furans(PCDFs), Toxaphene and other pollutants	http://www.ec.gc.ca/rs-mn/default.asp?lang=En&n=A31231AF-1
NDAMN – National Dioxin Air Monitoring Network	EPA	34	1998 - 2005	CDDs. CDFs, dioxin-like PCBs	http://cfpub.epa.gov/ncea/CFM/recordisplay.cfm?deid=54811
IADN -- Integrated Atmospheric Deposition Network	EPA	20	1990	PAHs, PCBs, and organochlorine compounds are measured in air and precipitation samples	http://www.epa.gov/greatlakes/monitoring/air2/
EMAP – Environmental Monitoring and Assessment Program	EPA	12,600	1988	Oriented to ecological and water monitoring	http://www.epa.gov/emap/index.html

APPENDIX G. FIELD CAMPAIGNS FOR NON-ROUTINE SPECIAL INTENSIVE STUDIES[3,4]

Network	Lead Agency[1]	Number of Sites	Initiated	Measurement Parameters	Location of Information and/or Data	Notes
Deriving Information on Surface Conditions from Column and Vertically Resolved Observations Relevant to Air Quality (DISCOVER-AQ)	NASA with MDE other organizations	3 aircraft 1 ship 6 MDE Surface air quality sites Ozonesondes Pandoras Dragon network	2011	Total column measurements of O_3, NO_2, CH_2O, and AOD. In situ profiles of O_3, NO_2, CH_2O, NO, NO_y, CO, CO_2, CH_4, H_2O, VOC, aerosol properties (size distribution, scattering, absorption, extinction, non-sphericity, f(RH), black carbon, soluble ions), and PBL height. Ground-based in situ measurements of O_3, NO_2, NO, NO_y, H_2O, $PM_{2.5}$, aerosol scattering. Ground- based lidar observations of aerosols and PBL height.	http://discover-aq.larc.nasa.gov/data. php	Deriving Information on Surface Conditions from Column and Vertically Resolved Observations Relevant to Air Quality (DISCOVER-AQ) is a NASA Earth Venture (EV) 1 mission. DISCOVER-AQ will conduct a series of field missions with an overarching objective to improve the interpretation of satellite observations to diagnose near-surface conditions relating to air quality and the following science objectives 1: Relate column observations to surface conditions for aerosols and key trace gases O_3, NO_2, and CH_2O; 2: Characterize differences in diurnal variation of surface and column observations for key trace gases and aerosols; 3: Examine horizontal scales of variability affecting satellites and model calculations. The first of the field missions was conducted over the greater metropolitan area of Baltimore, MD. With two NASA aircraft and an extensive ground network, DISCOVER- AQ operated as a complex observing system, providing multiple perspectives on the factors that control air quality and influence the ability to monitor pollution events from space. To

Network	Lead Agency[1]	Number of Sites	Initiated	Measurement Parameters	Location of Information and/or Data	Notes
						emulate a satellite observing system NASA's UC-12 King Air flew at an altitude of 27,000 feet with an UV/VIS spectrometer and High Spectral Resolution Lidar to collect remote sensing observations of particulate matter and gaseous pollutants. Detailed profile observations were collected underneath the UC-12 with the NASA's P- 3B using in situ sampling to provide detailed information on the distribution of pollutants from 1000 to 10,000 feet. These flights were anchored to a network of ground sites measuring surface air quality of particulate and trace gas pollutants, with additional profiles and total column measurements collected from the surface using tethered balloons, lidars, passive remote sensors, and ozonesondes. The DISCOVER-AQ mission will continue until 2014, with future deployments to California, San Joaquin Valley in January-February 2103, Houston, TX in September 2013, and a final deployment in Summer 2014 with the location to be determined.
CalNex 2010	NOAA with CARB	1 ship 4 aircraft 2 monitoring sites Ozonesonde	2010	Primary pollutants (CO, NO, NO2, SO2, NMHC, CO2, NH3, PM, VOC, black carbon, and	http://www.esrl.noaa.gov/csd/calnex/ whitepaper.pdf	This is a joint field study of atmospheric processes over California and the eastern Pacific coastal region that emphasizes the interactions between air quality and climate change issues, including those

(Continued)

Network	Lead Agency[1]	Number of Sites	Initiated	Measurement Parameters	Location of Information and/or Data	Notes
		network Radar wind profiler network		greenhouse gases); Secondary species: O_3, CH_2O, aldehydes, PAN, HNO_3, NO_3, N_2O_5, sulfuric acid, hydroxyl and peroxy radicals, aerosol size distribution and chemical composition; Other parameters (H_2O, aerosol properties, radiation, and meteorological parameters).		affecting the hydrologic cycle. The study constitutes one of a series of comprehensive regional air quality and climate assessments conducted by NOAA with expansion of CARB's leadership of California air quality studies. It complements the ongoing CEC regional climate change studies and brings together specialized, complementary resources such that the outcome will be able to address important scientific questions that have an impact on environmental policy. Airborne (NOAA WP-3D Orion, NOAA Twin Otter Remote Sensing Aircraft), ship (NOAA R/V Ronald H. Brown), on-going ground-based instrument packages (upper-air observations, ground-based chemical measurements), and satellite observations (MODIS, GOES) will be employed. The collaboration of agencies will link short-term data gathered during the field program to extensive surface observations, long term data sets, and California's advanced modeling capabilities for both regional air quality and climate.

Network	Lead Agency[1]	Number of Sites	Initiated	Measurement Parameters	Location of Information and/or Data	Notes
Arctic Research of the Composition of the Troposphere from Aircraft and Satellites (ARCTAS / POLARCAT)	NASA (with various universities & research institutions)	3 aircraft 1 monitoring site	2008 (spring / summer)	Primary pollutants (CO, NO, NO2, SO2, NMHC, CO2, NH3, PM, VOC, black carbon, and greenhouse gases); Secondary species: O3, CH2O, aldehydes, PAN, HN O3, N O3, N2O 5, sulfuric acid, hydroxyl and peroxy radicals, aerosol size distribution and chemical composition; Other parameters (H2O, aerosol properties, radiation, and meteorological parameters)	http://www-air.larc.nasa.gov/missions/arctas/arctas.html & data workshop	ARCTAS is a study of the impact of air pollution and forest fires on the arctic climate that integrates measurements from multiple aircraft and satellites. It has four major scientific themes: (1) long range transport of pollution to the Arctic including arctic haze, tropospheric ozone, and persistent pollutants such as mercury; (2) boreal forest fires and their implications for atmospheric composition and climate; (3) aerosol radiative forcing from arctic haze, boreal fires, surface deposited black carbon, and other perturbations; and (4) chemical processes with focus on ozone, aerosols, mercury, and halogens. ARCTAS is part of a larger interagency and international IPY effort collectively identified as POLARCAT which is intended to execute a series of aircraft experiments following pollution plumes as they are transported into the Arctic.
Aerosol, Radiation, and Cloud Processes Affecting Arctic Climate (ARCPAC)	NOAA (with University of Colorado, Georgia Institute of Technology, and NASA,	1 aircraft (NOAA WP-3D) in Fairbanks, Alaska	2008 (Spring)	Primary pollutants--CO, NO, NO_2, SO_2, CO_2, PM, VOCs, BC, GHGs; Secondary species--O3, PAN, HN O_3, sulfuric acid, BrO, BrCl, Cl2, Br2+ HOBr, aerosol size	http://www.esrl.noaa.gov/csd/arcpac/	ARCPAC is a field and modeling study of atmospheric chemical, microphysical, and radiative processes that affect Arctic climate. Particular foci of study are 1) the sources, transport and radiative effects of black carbon (soot) and non-absorbing particles in the Arctic, 2) the deposition of black carbon particles to the snow and sea-ice surface, and 3) the effects of

(Continued)

Network	Lead Agency[1]	Number of Sites	Initiated	Measurement Parameters	Location of Information and/or Data	Notes
	University of Alaska-Fairbanks)			distribution and chemical composition: Other parameters--H2O, aerosol optical properties, radiation, sea surface temperature, and meteorological parameters.		aerosol particles on cloud microphysical and radiative properties. Additional topics include the characteristics of halogen chemistry near the sea-ice surface, and the characteristics and atmospheric processing of pollutants emitted from anthropogenic sources at cold temperatures. The program includes a substantial effort to model the transport and transformations of pollutants in the Arctic, including assimilation of chemical data from satellite observations, and the effects of such pollutants on radiative transfer. ARCPAC is a participating campaign of the international POLARCAT program of the IPY, and includes substantial collaboration with the NASA ARCTAS and DOE ISDAC programs.
International Chemistry Experiment in the Arctic LOwer Troposphere (ICEALOT) --Organized as part of the International	NOAA (with various universities, research institutes, and a research corporation)	Research Vessel Knorr (Woods Hole Oceanographic Institute) with funding from NOAA /OAR Marine Charter Fund	2008 (March-April); Woods Hole, MA; Long Island Sound/NYC;	Aerosol measurements: light absorption, scattering, and extinction; particle concentration and size distributions; chemical composition; cloud nuclei	http://saga.pmel.noaa.gov/Field/icealot/	The ICEALOT field mission examined the transport and transformation of pollutants to and within the ice-free waters of the Arctic Ocean during springtime. The primary objectives were 1) to identify sources (local, regional, long-range) of pollutants in the marine boundary layer (MBL); 2) to examine the chemical evolution of aerosols and gases in the Arctic MBL,

Network	Lead Agency[1]	Number of Sites	Initiated	Measurement Parameters	Location of Information and/or Data	Notes
Polar Year (IPY), and a POLARCAT activity--			Tromso, Norway; Svalbard, Norway; Reykjavik Iceland	concentrations; vertical profiles of aerosol backscatter; AOD; Trace gas measurements: ANs, ClNO2, CO, CO2, DMS, H2O, NO, NO2, N O3, N2O 5 NOy, O3, PANs, POPs, Rn, SO2, VOCs; Meteorological and dynamical parameters		including photochemistry (both halogen and OH initiated), nocturnal chemistry, and physical transformations; and 3) to determine the impacts on the surface radiative balance from MBL aerosols and ozone, with specific focus on aerosol-cloud-radiative effects. A secondary objective of ICEALOT was to explore the wintertime chemical, photochemical, and physical processes associated with the evolution and transport of urban pollution into the North Atlantic MBL.
Texas Air Quality Study II (2005 - 2006)	Texas	17	2006	O3, NOx, NOy, SO2, Haze, Visibility, CO, VOC, Solar Radiation, Surface Meteorology, Upper Air	http://www.utexas.edu/research/ceer/texaqsII/PDF/12-12-04_Projected_Surface Sites_tbl.pdf	Researchers from universities, state and Federal agencies, private industry, and local governments joined forces to conduct a major field study to address air quality issues in the eastern half of Texas. The study, conducted in 2005 and 2006, examined regional ozone formation, transport of ozone and ozone precursors, meteorological and chemical modeling, issues related to ozone formation by highly reactive emissions, and particulate matter formation. The information from the study served as a scientific basis for State Implementation Plans (SIPs) for ozone (with concentrations averaged over 8 hours), regional haze, and fine

(Continued)

Network	Lead Agency[1]	Number of Sites	Initiated	Measurement Parameters	Location of Information and/or Data	Notes
						particulate matter (particulate matter less than 2.5 microns in diameter, $PM_{2.5}$
2006 Texas Air Quality Study/ Gulf of Mexico Atmospheric Composition and Climate Study (TexAQS/GoM ACCS)	NOAA	1 ship, 2 aircraft	2006	O3, NO, NO2, NOy, VOCs, CO2, CO, SO2, HN O3, NH3, other reactive pollutants, aerosols, meteorological parameters & upper air	http://esrl.noaa.gov/c sd/2006/	For TexAQS 2006, the NOAA air quality component investigated, through airborne and sea-based measurements, the sources and processes that are responsible for photochemical pollution and regional haze during the summertime in Texas. The focus of the study was the transport of ozone and ozone precursors within the state and the impact of the long-range transport of ozone or its precursors.
Intercontinental Chemical Transport Experiment - North America (INTEX-B) -- Intercontinental Transport and Chemical Transformation (ITCT/IGAC)	NOAA	3 aircraft	2006	O3, NO, NO2, NOy, VOCs, CO2, CO, SO2, HN O3, NH3, other reactive pollutants, aerosols, meteorological parameters, altitude -- NOAA aircraft	http://cloud1.arc.nasa.gov/intex-b/	The export of air pollutants from urban to regional and global environments is a major concern because of wide-ranging potential consequences for human health, cultivated and natural ecosystems, visibility degradation, weather modification, changes in radiative forcing, and tropospheric oxidizing capacity. During the spring of 2006 a highly integrated atmospheric field experiment was performed over and around North America. The Megacity Initiative: Local and Global Research Observations (MILAGRO), http://www.eol.ucar.edu/projects/milagro/, resulted through a highly coordinated collaboration between NSF

Network	Lead Agency[1]	Number of Sites	Initiated	Measurement Parameters	Location of Information and/or Data	Notes
						(through MIRAGE-Mex), DOE (through MAX-Mex), NASA (through INTEX-B) and a variety of research institution in the U.S. and Mexico and involved ground and air borne activities centered on Mexico City, Mexico during March 2006. MILAGRO goals were greatly facilitated and enhanced by a number of concurrent and coordinated national and international field campaigns and global satellite observations. After MILAGRO, NASA continued investigating this issue, this time focusing on the influence of Asian pollutants on North America, through a second airborne field element of INTEX-B in collaboration with NSF and NCAR. The integrated goals of MILAGRO and INTEX are: • To study the extent, persistence, and transformation of Mexico City pollution plumes; • To relate atmospheric composition to sources and sinks; • To quantify radiative properties and effects of aerosols, clouds, water vapor & surfaces; • To map anthopogenic and biogenic emissions; • To characterize transport and evolution of Asian pollution to North America

(Continued)

Network	Lead Agency[1]	Number of Sites	Initiated	Measurement Parameters	Location of Information and/or Data	Notes
						and beyond and determine implications for regional air quality and climate; • To achieve science-based validation of satellite observations of tropospheric composition
International Consortium for Atmospheric Research on Transport and Transformation (ICARTT)	NOAA (with various other agencies and research institutions)	Multiple aircraft and other measurement platforms	2004	Surface sites and networks, mobile platforms (aircraft and ship) and satellite data were used for measurement parameters; see http://www.esrl.noaa.gov/csd/ICARTT/fieldoperations/ for detailed information	http://www.esrl.noaa.gov/csd/ICARTT/index.shtml	ICARTT was formed to study the sources, sinks, chemical transformations and transport of ozone, aerosols and their precursors to, and over, the North Atlantic Ocean. Groups in North America and Europe had independently developed plans for field experiments in the summer of 2004 that shared many of the same goals and objectives in overlapping study areas; the plans were aimed at developing a better understanding of the factors that shape air quality in their respective countries and the remote regions of the North Atlantic. ICARTT was formed to take advantage of this synergy by planning and executing a series of coordinated experiments to study the emissions of aerosol and ozone precursors and their chemical transformations and removal during transport to, and over, the North Atlantic. The combined research conducted in the programs that make up ICARTT focus on regional air quality, intercontinental

Network	Lead Agency[1]	Number of Sites	Initiated	Measurement Parameters	Location of Information and/or Data	Notes
						transport, and radiation balance in the atmosphere.
Intercontinental Chemical Transport Experiment - North America (INTEX-NA) -- Intercontinental Transport and Chemical Transformation (ITCT/IGAC)	NASA	aircraft, sondes, satellites	2004	O3, NO, NO2, NOy, VOCs, CO2, CO, SO2, HN O3, NH3, other reactive pollutants, aerosols, meteorological parameters, altitude -- NOAA aircraft	http://cloud1.arc.nasa.gov/intex-na/desc.html	INTEX-NA is an integrated atmospheric field experiment performed over and around North America. It seeks to understand the transport and transformation of gases and aerosols on rans ontinental/intercontinental scales and their impact on air quality and climate. A particular focus in this study is to quantify and characterize the inflow and outflow of pollution over North America. The main constituents of interest are ozone and precursors, aerosols and precursors, and the long-lived greenhouse gases. INTEXNA is part of a larger international ITCT (Intercontinental Transport and Chemical Transformation) initiative. INTEX-NA goals are greatly facilitated and enhanced by a number of concurrent and coordinated national and international field campaigns and satellite observa-tions. Synthesis of the ensemble of observations from surface, airborne, and space platforms, with the help of a hierarchy of models is an important goal of INTEX-NA.

(Continued)

Network	Lead Agency[1]	Number of Sites	Initiated	Measurement Parameters	Location of Information and/or Data	Notes
New England Air Quality Study (NEAQS) -- Intercontinental Transport and Chemical Transformation (ITCT) 2004	NOAA	4 site, 1 ship, 2 aircraft, profiler network	2004	O3, NO, NO2, NOy, VOCs, CO2, CO, SO2, HN O3, NH3, other reactive pollutants, aerosols, meteorological parameters & upper air	http://esrl.noaa.gov/csd/2004/	NOAA continues a joint regional air quality and climate change study combining elements of the previous NEAQS study and the Intercontinental Transport and Chemical Transformation (ITCT) research activity to focus on air quality along the Eastern Seaboard and transport of North American emissions into the North Atlantic. The major NOAA assets (the two aircraft and the ship) are deployed in a manner that supports the objectives of both components.
East Tennessee Ozone Study (ETOS)	NOAA	15+	2003	O3, Surface Meteorology	http://www.arl.noaa.gov/etos_122005.php	ETOS 2003 developed a regional ozone database to include both mean hourly averages and hourly histograms of individual measurement readings. The 2003 study period (based on scoping studies 1999 - 2002) provides a regional view to supplement Tennessee's regulatory network and serves as a demonstration and evaluation/validation database for various operational and developmental air quality forecast model components. The full scope of ETOS 2000 is continuously under planning and review, and is refined each year using the previous year's analysis and experience

Network	Lead Agency[1]	Number of Sites	Initiated	Measurement Parameters	Location of Information and/or Data	Notes
						to focus on particular issues within the East Tennessee region.
Texas Air Quality Study (TexAQS) 2000	Texas	~20	2002	O3, NOx, PM2.5/ PM10, CO, SO2, VOCs, Surface Meteorology	http://www.utexas.edu/research/ceer/texaqs/visitors/about.html	The study is designed to improve understanding of the factors that control the formation and transport of air pollutants along the Gulf Coast of southeastern Texas. Six weeks of intensive sampling, including measurements of gaseous, particulate, and hazardous air pollutants, are made at approximately 20 ground stations, located throughout the eastern half of the state. Experts in meteorology, atmospheric chemistry, and other areas of science study the formation, composition, and day-night cycles of ozone and particulate matter, as well as how these pollutants are affected by weather.
Texas Air Quality Study (TexAQS) 2000 Field Campaign	NOAA	2 aircraft	2002	O3, CO, CO2, SO2, NO, NO2, NOy, PAN, HN O3, NH3, VOCs, Solar Radiation, Meteorological Parameters, aerosols	http://www.utexas.edu/research/ceer/texaqs/visitors/about.html	Additional sampling in TexAQS 2000 is carried out with specially equipped aircraft that can detect air pollutants very quickly, at very low concentrations.
Bay Region Atmospheric Chemistry Experiment (BRACE)	NOAA	1 aircraft	2002	N O3, NH4, O3, SO2, NOx, CO, trace metals, particulates	http://hsc.usf.edu/publichealth/EOH/BRACE/	The Florida Department of Environmental Protection (FDEP), with the support of a team of Federal, state, local, university and private scientists (including NOAA), conducted a month-long series of intensive studies to

(Continued)

Network	Lead Agency[1]	Number of Sites	Initiated	Measurement Parameters	Location of Information and/or Data	Notes
						determine the level of influence of nitrogen deposited into Tampa Bay from local and regional sources of air pollutants on water quality. During the Bay Region Atmospheric Chemistry Experiment (BRACE), NOAA operated a research aircraft over the Tampa Bay region to collect air quality measurements of the many atmospheric forms of nitrogen and related pollutants that may potentially influence the water quality of Tampa Bay.
New England Air Quality Study (NEAQS) 2002 -- AIRMAP	NOAA	4	2002	O_3, NO_x, NO_y, SO_2, CO, VOCs, $PM_{2.5}$, Precipitation Chemistry, Surface Meteorology	http://airmap.unh.edu/data/	AIRMAP is a research program focused on atmospheric chemical and physical observations in rural to semi-remote areas of New Hampshire with the goal of understanding inter-relationships in regional air quality, meteorology, and climatic phenomena. Research goals are to: (1) document and analyze current trends in the regional air quality of New England which is affected by transport from upwind regions of the U.S. and Canada and by local emission sources; (2) document and analyze current and past (the last 100 years) synoptic-to-local meteorological patterns, features, and extreme events in New England; and (3) numerically simulate the coupled

Network	Lead Agency[1]	Number of Sites	Initiated	Measurement Parameters	Location of Information and/or Data	Notes
						evolution of atmospheric transport and chemistry in New England using various modeling tools.
New England Air Quality Study (NEAQS) 2002	NOAA	1 ship, 2 aircraft	2002	O3, NO, NO2, NOy, VOCs, CO2, CO, SO2, HN O3, NH3, other reactive pollutants, aerosols, meteorological parameters & upper air	http://esrl.noaa.gov/csd/NEAQS/	The NOAA component of this multi-institutional effort addresses the analysis of existing climate data, and the development of new air quality monitoring programs. A background of information is to be developed that addresses New England's changing climate and air quality so as to improve understanding of the relationship between air quality and weather and determine the causes of climate change in New England
Intercontinental Transport and Chemical Transformation (ITCT) 2002 Activities	NOAA	1 site, 1 aircraft	2002	CO2, CO, CH4, SO2, O3, SF6, N2O, CFCs, Aerosols, Solar Radiation, Surface Meteorology & Upper Air -- surface. O3, NO, NO2, NOy, VOCs, CO2, CO, SO2, HN O3, NH3, other reactive pollutants, aerosols, meteorology-cal parameters & upper air – aircraft	http://esrl.noaa.gov/csd/ITCT/2k2/	This field program in spring 2002 investigated the composition of air masses along the Pacific coast of North America, is part of the Intercontinental Transport and Chemical Transformation (ITCT) research activity of the International Global Atmospheric Chemistry Program (IGAC) Program. Goals of this field study are to: characterize the chemical composition of the air masses coming ashore at the West Coast; explore the composition of these air masses as they are transported inland; and investigate the alteration in composition associated with the addition of emissions from U.S. West Coast sources. The NOAA WP-3D aircraft

(Continued)

Network	Lead Agency[1]	Number of Sites	Initiated	Measurement Parameters	Location of Information and/or Data	Notes
						deployed a wide array of instrumentation for the in situ measurement of gaseous and aerosol parameters plus radiation and remote aerosol sensing by LIDAR. The Trinidad Head baseline observatory characterizes chemical composition of marine boundary layer at the U.S. West Coast and provides linkage between composition measurements and radiative properties of the aerosols. The NOAA ETL Laboratory network of 915-MHz radar wind profilers that are deployed in California provide additional meteorological information.
TRAnsport and Chemical Evolution over the Pacific (TRACE-P)	NASA	2 aircraft	2001 (2 months)	O3, NO, NO2, N2O, CH4, SO2, NH3, CO, CO2, aerosols, PAN, HN O3, aldehydes, peroxides, speciated hydrocarbons, other pollutants, meteorological parameters	http://www-gte.larc.nasa.gov/gte_fld.htm#TRACE	TRACE-P is part of a series of aircraft missions aimed at better understanding of global tropospheric chemistry, and more specifically in this case, the effects of outflow from the Asian continent on the composition of the global atmosphere. Objectives are to determine: (1) pathways for outflow of chemically and radiatively important gases and aerosols, and their precursors, from eastern Asia to the western Pacific; and (2) the chemical evolution of the Asian outflow over the western Pacific, and the ensemble of processes that control this evolution. Approximately 20 aircraft measurement

Network	Lead Agency[1]	Number of Sites	Initiated	Measurement Parameters	Location of Information and/or Data	Notes
						flights involving horizontal and vertical profiles for a total of over 300 hours were supported by surface-based measurements and soundings.
Aerosol Characterization Experiments - Asia (ACE-Asia)	NSF	sites, ships, aircraft, satellites	2001 (spring)	aerosol chemical, physical, and radiative properties and radiative fluxes, meteorological parameters	http://saga.pmel.noaa.gov/Field/aceasia/ACEAsiaDescription.html	The Aerosol Characterization Experiments (ACE) are designed to increase understanding of how atmospheric aerosol particles affect the Earth's climate system. ACE-Asia took place during the spring of 2001 off the coast of China, Japan and Korea which includes many types of aerosol particles of widely varying composition and size. These particles include those emitted by human activities and industrial sources, as well as wind-blown dust. Data from ACE-Asia is improving understanding of how atmospheric aerosols influence the chemical and radiative properties of the Earth's atmosphere.
Central California Ozone Study (CCOS)[2]	California	100+ sites, 6 aircraft, profilers, sondes	2000	O3, VOC, NOx, NO, NOy, CO, PM10, PM2.5, solar radiation, surface meteorology, upper air	http://www.arb.ca.gov/airways/	For the summer season, this study collected meteorological and air quality data for the central section of California in 2000. Planes and weather balloons collected data at ground level and aloft. The data collected is used to improve the understanding of the role of meteorology on the formation and behavior of air pollutants and their precursors and emission sources and patterns. The information gathered will be used to develop an improved modeling system

(Continued)

Network	Lead Agency[1]	Number of Sites	Initiated	Measurement Parameters	Location of Information and/or Data	Notes
						that will be used in preparing plans to attain the new federal 8-hour ozone standard, as well as to update the Clean Air Plan to attain the state ozone standard.
California Regional Particulate Air Quality Study (CRPAQS)[2]	California	~60	1999 to 2001	PM2.5, PM10, nephelometer, with some sites adding SO4/N O3, OC/EC, NO2, NOy, PAN, SO2, surface meteorology	http://www.arb.ca.gov/airways/	The California Regional PM10/PM2.5 Air Quality Study is a comprehensive public/private sector collaborative program to provide an improved understanding of particulate matter and visibility in central California. It is intended to evaluate both the national and State air quality standards for PM10 and PM2.5. The field programs consisted of 14 months of monitoring throughout the San Joaquin Valley (SJV) and surrounding regions, as well as intensive monitoring during summer, fall, and winter seasonal periods.
Southern Oxidant Study (SOS) 1999 Field Campaign - - Nashville	NOAA	3 sites, 4 aircraft	1999	O3, NO, NO2, NOy, VOCs, aerosols, Surface Meteorology & Upper Air (profiler), ozonesonde – Surface O3, NO, NO2, NOy, VOCs, CO2, CO, SO2, HN O3, NH3, other	http://esrl.noaa.gov/csd/SOS99/	The Southern Oxidants Study (SOS), in collaboration with other organizations and programs, conducted this major Field Campaign during June/July 1999. The Nashville/Middle Tennessee region measurements focused on an improved understanding of the processes that control the formation and distribution of fine particles and ozone. Three study themes were: local vs. regional contrasts,

Network	Lead Agency[1]	Number of Sites	Initiated	Measurement Parameters	Location of Information and/or Data	Notes
				reactive pollutants,aerosols, meteorolo-gical parameters, altitude -- Aircraft		ozone and PM formation in plumes, and diurnal cycle in chemistry and meteorology. These themes were addressed through a series of coordinated measurements involving instrumented aircraft and a ground-based network of chemistry and meteorological measurements.
PM Supersite Program	EPA	2 Phase I Sites 7 Phase II Sites	1999	Measurement may include: PM2.5, PM10, TSP, SO_4, NO_3, EC, OC, light absorption & extinction, O_3, CO , NOx, NO, NO_2, NOy, HNO_3, NH_3, VOCs, Carbonyls, PAH, major ions and elements, surface and upper air meteorology	http://www.epa.gov/ttn/amtic/supersites.html	A "Supersites Conceptual Plan" was developed and implemented in response to Executive and Congressional mandates and recommendations from the National Research Council. Atlanta and Fresno were selected as initial Phase I sites and, as a result of a competitive process, Baltimore, Fresno, Houston, Los Angeles, New York, Pittsburgh, and St. Louis were selected for Phase II. Goals generally were to characterize particulate matter, support health effects and exposure research, and conduct methods testing. Extensive monitoring, data analysis, and publication continued to 2005 with the preparation of a Final Report for each city.
Big Bend Regional Aerosol and Visibility Observational (BRAVO) Study	NPS/EPA	38 fixed, 6 tracer sites	1999	SO_2, SO_4, PM2.5, NO_3, NH_4, major ions and elements, nephelometer, transmissometer, meteorological	http://vista.cira.colostate.edu/improve/studies/BRAVO/reports/projectplan/BRAVOStudyPlan.htm	The BRAVO study was conducted for four months during 1999 with the primary objective of identifying the causes of haze in the Big Bend National Park located in West Texas. This very large, collaborative study enlisted numerous participants with sponsorship

(Continued)

Network	Lead Agency[1]	Number of Sites	Initiated	Measurement Parameters	Location of Information and/or Data	Notes
				parameters & upper air, PFC tracer	http://www.nature.nps.gov/air/studies/bravo/index.cfm	from Federal/state agencies, private industry, and research organizations. The BRAVO study utilized data from a 38-site network to characterize spatial and temporal aerosol patterns in the atmosphere. In addition, upper-air measurements and extensive optical measurements of light scattering and absorption were made. Because monitoring and source characterization activities were conducted only in the United States, the study design included additional monitoring and tracer studies along the U.S./Mexican border.
Indian Ocean Experiment (INDOEX)	UCSD	6 sites, 2 ships, 5 aircraft, satellites	1999 (4 months)	O3, NO, NO2, VOCs, CO2, CO, SO2, HN O3, NH3, other reactive pollutants, trace gases, aerosols, meteorological parameters & upper air	http://data.eol.ucar.edu/codiac/projs?INDO EX	The Indian Ocean Experiment (INDOEX) addresses questions of climate change through collection of in-situ data on the regional cooling effect of sulfate and other aerosols. The project's goal is to study natural and anthropogenic climate forcing by aerosols and feedbacks on regional and global climate. INDOEX field studies occur where pristine air masses from the southern Indian Ocean including Antarctica and not-so-clean air from the Indian subcontinent meet over the tropical Indian Ocean to provide a unique natural laboratory for studying aerosols. Scientists collect data from the

Network	Lead Agency[1]	Number of Sites	Initiated	Measurement Parameters	Location of Information and/or Data	Notes
						water surface through the lower stratosphere, on the aerosol composition, reactive atmospheric gases, solar radiation fluxes, wind and water vapor distribution. To this end, investigators use multiple aircraft, ships and island stations over the Arabian Sea and the Indian Ocean.
Eulerian Model Evaluation Field Study (EMEFS)	Canada	~135	1998	O3, NO2, SO2, NH3, HN O3, majorions,	http://www.ec.gc.ca/natchem/default.asp?lang=en&n=9E05341A- 1	Under EMEFS, air and precipitation chemistry data were collected daily for two years over much of the eastern United States and Canada to provide data for assessing the performance of acid deposition and other regional scale models.
NARSTO-Northeast 1995	Multiple	559	1995	O3, NO, NOx	http://eosweb.larc.nasa.gov/GUIDE/dataset_documents/narsto_ne_model.html	Measurements were made during the NARSTO-Northeast 1995 intensive field campaign during the period May through September. One-hour average O3, NO, and NOX measurement results are reported for ground surface monitoring stations operated by various agencies including EPA AIRS, CASTNet, ESE, Harvard University, NYSEG, PEPCO, and the University of Maryland.
SOS Nashville/Middle Tennessee Ozone Study	TVA	116	1994-1995	O3, SO2, NO, NOy, and CO, VOC, Surface Meteoro-logy, rawinsonde and ozonesonde releases, and a radar	http://www.ncsu.edu/sos/pubs/sos2/State_of_SOS_2.pdf	This ozone-focused field study was carried out in the 11-state region surrounding Nashville/Middle Tennessee, beginning with a 3-week exploratory study during the summer of 1994 and culminating in a six-week field

(Continued)

Network	Lead Agency[1]	Number of Sites	Initiated	Measurement Parameters	Location of Information and/or Data	Notes
				profiler/radar acoustic sounding system. -- surface Airborne ozone and aerosol lidar – aircraft		measurement campaign June/July 1995. Measurements were taken at 116 ground-based and tall building and tower based chemical and meteorological measurement sites and a series of six airborne chemical measurement platforms. The most significant feature of the Nashville/Middle Tennessee Ozone Study was a coordinated series of 40+ aircraft studies to measure physical and chemical characteristics of urban and industrial plumes. (Note: an earlier ozone-focused set of field studies was also conducted in the Atlanta, GA area during the summers of 1990 - 1992.)
North Atlantic Regional Experiment (NARE)	NOAA	various sites, 1 ship	1993, 1996, 1997	O3, NO, NO2, NOx, NOy, VOC, Surface Meteorology	http://www. igacproject.org/ Newsletters (Issue24/August 2001)	The NARE program measured the type and amount of air pollutants being transported from the North American continent to the Northern Atlantic Ocean. Since the Northeast United States and Nova Scotia, Canada, are the last land locations as air masses move out over the ocean, measurements were made a number of land and island sites in Maine, Nova Scotia, and Sable Island. Acadia National Park participated in this study.

Footnotes:

1. EPA -- Environmental Protection Agency

MDE -- Maryland Department of Environment
NASA -- National Aeronautics and Space Administration
NOAA -- National Oceanic and Atmospheric Administration
NPS -- National Park Service
NSF -- National Science Foundation
CARB -- California Air Resources Board
CEC -- California Energy Commission
UCSD -- University of California San Diego (Scripps Institution of Oceanography)

2. This study is part of the Central California Air Quality Studies (CCAQS) which comprise the California Regional Particulate Air Quality Study (CRPAQS) and the Central California Ozone Study (CCOS). CCAQS is a multi-year effort of meteorological and air quality monitoring, emission inventory development, data analysis, and air quality simulation modeling. Prior studies in California included: Southern California Ozone Study (SCOS97) -- 1997; Integrated Monitoring Study (IMS95) -- 1995; San Joaquin Valley Air Quality Study (SJVAQS) -- 1990; SARMAP Ozone Study -- 1990; Southern California Air Quality Study (SCAQS) -- 1987.
3. Historically, there have been many other field studies in the 1960's - 1990's that are not reflected in this table that involve both fixed monitoring sites and aircraft; well known examples include Regional Air Pollution Study (RAPS), Large Power Plant Effluent Study (LAPPES), Northeast Corridor Regional Modeling Program (NECRMP), Northeast Regional Oxidant Study (NEROS), Persistent Elevated Pollutant Episode (PEPE), and Lake Michigan Ozone Study (LMOS).
4. In addition to the air monitoring networks and related studies detailed in this table that are primarily concerned with lower tropospheric air pollution, there are a large number of observations and studies conducted by NASA, NOAA, and others that address such topics as (1) upper tropospheric and stratospheric ozone and aerosols, (2) cloud processes, and (3) validation experiments for satellite observations. These studies include but are not limited to:
   - Stratospheric Tropospheric Exchange Project (STEP) – 1987
   - Airborne Antarctic Ozone Experiment (AAOE) – 1987
   - Airborne Arctic Stratospheric Experiment (AASE) – 1989
   - Airborne Arctic Stratospheric Experiment II (AASE2) – 1992
   - Stratospheric Photochemistry Aerosols and Dynamics Experiment (SPADE) – 1993
   - Airborne Southern Hemisphere Ozone Experiment / Measurements for Assessing the Effects of Stratospheric Aircraft (ASHOE/MAESA) – 1994

- Stratospheric Tracers of Atmospheric Transport (STRAT) – 1995-1996
- Tropical Ozone Transport Experiment (TOTE) and Vortex Ozone Transport Experiment (VOTE) – 1995-1996
- Subsonic Aircraft: Contrail and Clouds Effects Special Study (SUCCESS) – 1996
- Photochemistry and Ozone Loss in the Arctic Region in Summer (POLARIS) – 1997
- Subsonic Assessment: Ozone and Nitrogen Oxide Experiment (SONEX) – 1997
- Texas Florida Underflights A (TEFLUN) – 1998
- The Third Convection and Moisture Experiment (CAMEX 3) – 1998
- TRMM Brazil Validation Experiment (TRMM-LBA) – 1999
- TRMM Kwajalein Validation Experiment (KWAJEX) – 1999
- Nauru 1999 Field Campaign – 1999
- South African Fire-Atmosphere Research Initiative 2000 (SAFARI) – 2000
- SAGE III Ozone Loss and Validation Experiment (SOLVE) – 1999-2000
- ERAST Predator-B RPV Homepage (ERAST) – 2000
- CAMEX 4 The Fourth Convection and Moisture Experiment (CAMEX 4) – 2001
- East Pacific Investigation of Climate (EPIC) 2001 Field Program – 2001
- The Cirrus Regional Study of Tropical Anvils and Cirrus Layers-Florida Area Cirrus Experiment (CRYSTAL FACE) – 2002
- The SAGE III Ozone Loss and Validation Experiment (SOLVE II) – 2003
- The Aura Validation Experiment (AVE) – 2004
- The Intercontinental Chemical Transport Experiment – North America (INTEX-NA) – 2004
- The Aura Validation Experiment Houston (AVE Houston) – 2004
- North American Monsoon Experiment (NAME) – 2004
- Winter Storms Reconnaissance Program 2004 (WSR2004) – 2004
- Polar Aura Validation Experiment (PAVE) – 2005
- The Tropical Cloud Systems and Processes Mission (TCSP) – 2005
- UAS Flight Demonstration Project 2005 – 2005

APPENDIX H. SATELLITE – BASED AIR QUALITY OBSERVING SYSTEMS[1]

Instrument	Satellite Platform[4]	Lead Agency	Initiated /Planned	Measurement Parameters[2]	Orbit & Horizontal Resolution	Location of Information and/or Data
OLS (Operational Linescan System)	DMSP satellites	DOD	1962?	Identify fires and smoke plumes	Polar Imagery only	http://www.losangeles.af.mil/library/factsheets/factsheet.asp?id=5317
BUV (Backscatter Ultraviolet Spectrometer)	Nimbus 4	NASA	1970-1980	O3, CO2, SO2	Sun synchronous	http://nssdc.gsfc.nasa.gov/earth/nimbus_sensor.html
SBUV (Solar Backscatter Ultraviolet Spectrometer)	Nimbus 7	NASA	1978-1993	O3, SO2	Polar	http://nssdc.gsfc.nasa.gov/earth/nimbus_sensor.html
TOMS (Total Ozone Mapping Spectrometer)	Nimbus 7 Meteor 3 Earth-Probe	NASA	1978-1993 1991-1994 1996-2005	O3, SO2, Aerosols	Polar $\sim 100km^2$	http://nssdc.gsfc.nasa.gov/earth/nimbus_sensor.html
LIMS (Limb Infrared Monitor of the Stratosphere)	Nimbus 7	NASA	1978-1979	O3, HN O3, NO2,	Polar	hhttp://nssdc.gsfc.nasa.gov/earth/nimbussensor.html
ATMOS (Atmospheric Trace Molecule Spectroscopy)	Spacelab 3 ATLAS -- 1,2,3	NASA	1985, 1992, 1993, 1994	O3, CFCl3, CF2Cl2, ClONO2, HCl, HF, CO, CH4, HCN, HN O3, NO, NO2, N2O, N2O5, Aerosols		http://asd-www.larc.nasa.gov/spectroscopy/ASDatmos.html
CLAES (Cryogenic Limb Array Etalon Spectrometer)	UARS	NASA	1991-1993	O3, CFCl3, CF2Cl2 ClONO2, CH4, HN O3, NO, NO2, N2O, N2O 5, Aerosols		http://umpgal.gsfc.nasa.gov/
HALOE (Halogen Occultation Experiment)	UARS	NASA	1991-2005	O3, HCl, HF, CH4, NO, NO2, Aerosols		http://umpgal.gsfc.nasa.gov/

(Continued)

Instrument	Satellite Platform[4]	Lead Agency	Initiated /Planned	Measurement Parameters[2]	Orbit & Horizontal Resolution	Location of Information and/or Data
ISAMS (Improved Stratospheric and Mesospheric Sounder)	UARS	NASA	1991-1992	O3, CO, CH4, NO2, N2O, N2O 5, Aerosols		http://umpgal.gsfc. nasa.gov/
MLS (Microwave Limb Sounder)	UARS	NASA	1991-1999	O3, ClO, CH3CN, HN O3, SO2		http://umpgal.gsfc. nasa.gov/
GOES Imager (Geostationary Operational Environmental Satellites)	GOES-13 GOES-15	NOAA	1994	Fire products for WF_ABBA, HMS (imagery) and GASP (aerosol optical depth)	Geostationary 4x4 km^2	http://cimss.ssec.wisc.e du/goes/da ta.html
GOES Sounder (Geostationary Operational Environmental Satellites)	GOES-13 GOES-15	NOAA	1994	Total column O3	Geostationary	http://cimss.ssec.wisc.e du/goes/da ta.html
AVHRR[3] (Advanced Very High Resolution Radiometer) AVHRR/3	NOAA-15 NOAA-16 NOAA-17 NOAA-18 NOAA-19 MetOp-B	NOAA EUMET SAT ESA	1998 2012	Aerosol optical depth, particle size information (over ocean) and vegetation/drought index products related to air quality through fires, and thermal fire products	Polar ~1x1 km^2	http://noaasis.no aa.gov/NOAASIS/ ml/avhrr.html http://www.esa.in t/esaLP/SEMV68L 8IOE_LPmetop_0. html
SBUV/2[3] (Solar Backscattered Ultraviolet Radiometer Model 2)	NOAA-16 NOAA-17 NOAA-18 NOAA-19	NOAA	2000	Total and profile O3 from surface to top of atmosphere in ~5 km thick Umkehr layers	Polar	http://www.ozonelayer. noaa.gov/action/ sbuv2.htm

Instrument	Satellite Platform[4]	Lead Agency	Initiated /Planned	Measurement Parameters[2]	Orbit & Horizontal Resolution	Location of Information and/or Data
MOPITT (Measurement of Pollution in the Troposphere)	EOS Terra	NASA	1999	CO, CH4	Polar 22 x 22 km^2	http://www.eos.ucar.edu/mopitt/
MISR (Multi-angle Imaging SpectroRadiomenter)	EOS Terra	NASA	1999	Aerosol properties and plume height information near the vicinity of fires	Polar ~1x1 km^2	http://misr.jpl.nasa.gov/
MODIS (Moderate Resolution Imaging Spectroradiometer)	EOS Terra EOS Aqua[5]	NASA	1999 2002	O3, Aerosol optical depth, particle size information, fine particle fraction, and thermal fire products	Polar ~1x1 km^2	http://modarch.gsfc.nasa.gov/index.p hp
AIRS (Atmospheric Infrared Sounder)	EOS Aqua[5]	NASA	2002	Total column ozone, surface temperature, temperature and moisture vertical profiles, (plus under development are CO and CO2 total column, O3 vertical distribution, and CH4 distribution)	Polar 50km	http://airs.jpl.nasa.gov/
ACE-FTS (Atmospheric Chemistry Experiment – Fourier Transform Spectrometer)	SCISAT-1	Canadian Space Agency	2003	O3, CO,SO2, PAN, H2O, CO2, CH4, CFC- 11, CFC-12, N2O, NH O3, NH3, C2H4, CH3OH, HCOOH	Polar	http://www.ace.uwaterloo.ca/index.html
HIRDLS (High Resolution Dynamics Limb Sounder)	EOS Aura[5]	NASA	2004	O3, CFC11, CFC12, ClONO2, CH4, HN O3, NO2, N2O, N2O 5, Aerosols	Polar	http://aura.gsfc.nasa.gov/index.html http://www.nasa.gov/mission_pages/aura/spacecraft/index.html
MLS (Microwave Limb Sounder)	EOS Aura[5]	NASA	2004	O3, BrO, ClO, HOCl, HCl, CO, HCN, CH3CN, HN O3, N2O, OH, HO2	Polar	http://aura.gsfc.nasa.gov/index.html

(Continued)

Instrument	Satellite Platform[4]	Lead Agency	Initiated /Planned	Measurement Parameters[2]	Orbit & Horizontal Resolution	Location of Information and/or Data
						http://www.nasa.gov/ mission_pages/aura/ spacecraft/index.html
OMI (Ozone Monitoring Instrument)	EOS Aura[5]	NASA	2004	O3, BrO, OClO, HCHO, CHOCHO, NO2, SO2 and absorbing aerosols	Polar 48 x 48 km2	http://aura.gsfc.nasa. gov/index.html http://www.nasa.gov/ mission_pages/aura/ spacecraft/index.html
TES (Total Emission Spectrometer)	EOS Aura[5]	NASA	2004	O3, CO, CH4, HN O3, NH3	Polar 26 x 42 km^2	http://aura.gsfc.nasa. gov/index.html http://www.nasa.gov/ mission_pages/aura/ spacecraft/ind ex.html
CALIOP (Cloud Aerosol Lidar with Orthogonal Polarization)	CALIPSO[5]	NASA	2005	Total aerosol backscatter, aerosol optical depth, and extinction	Polar 0.3 x 0.3 km^2	http://www-calipso. larc.nasa.go v/about/
OMPS (Ozone Mapping and Profiling Suite)	Suomi NPP (Suomi National Polar-orbiting Partnership) ,JPSS-1 (Joint Polar Satellite System) JPSS-2	NOAA, NASA, EUMET SAT	2011 2017 2023	Total column and vertical profile ozone data, aerosol index, aerosol optical depth, and SO2	Polar	http://jointmission.gsfc. nasa.gov/spacecraft_in st.html http://www.nesdis.noaa .gov/jpss/

Instrument	Satellite Platform[4]	Lead Agency	Initiated /Planned	Measurement Parameters[2]	Orbit & Horizontal Resolution	Location of Information and/or Data
VIIRS (Visible Infrared Imaging Radiometer Suite)	Suomi NPP (Suomi National Polar-orbiting Partnership), JPSS-1 (Joint Polar Satellite System) JPSS-2	NOAA, NASA, EUMETSAT	2011 2017 2023	Aerosol optical depth, suspended matter, aerosol particle size , and thermal fire products	Polar 8 x 8 km^2	http://jointmission.gsfc.nasa.gov/spacecraft_inst.html http://www.nesdis.noaa.gov/jpss/
CrIS (Cross Track Infrared Sounder)	Suomi NPP (Suomi National Polar-orbiting Partnership), JPSS-1 (Joint Polar Satellite System) JPSS-2	NOAA, NASA, EUMETSAT	2011 2017 2023	CO, CH4, and CO2	Polar 50 x 50 km^2	http://jointmission. gsfc.nasa.gov/spacecraft_inst.html http://www.nesdis.noaa.gov/jpss/
OCO-2 (Orbiting Carbon Observatory-2)	OCO-2[5]	NASA	2014 (planned)	CO2	Polar	http://oco.jpl.nasa. gov/
SAGE-III (Stratospheric Aerosol and Gas Experiment)	ISS (International Space Station)	NASA	2015	Aerosol extinction, ozone, water vapor, nitrogen dioxide, nitrogen trioxide, chlorine dioxide, in the mesosphere, stratosphere, and upper troposphere with a vertical resolution of 0.5 - 1 km resolution.	Inclined, non-sun-synchronos	http://www.nasa.gov/mission_pages/station/research/experiments/1004.html

(Continued)

Instrument	Satellite Platform[4]	Lead Agency	Initiated /Planned	Measurement Parameters[2]	Orbit & Horizontal Resolution	Location of Information and/or Data
ABI (Advanced Baseline Imager)	GOES-R (Geostationary Operational Environmental Satellite) GOES-S	NOAA	2016 2017	Aerosol optical depth, thermal fire products	Geostationary 2 x2 km^2	http://www.goes-r.gov/
GLM (Geostationary Lightening Mapper)	GOES-R (Geo-stationary Operational Environmental Satellite) GOES-S	NOAA	2016 2017	Total lightening, (in-Cloud, Cloud-Cloud, Cloud-Ground) detection for NO parameterization	Geostationary	http://www.goes-r.gov/
TEMPO UV-Vis Imaging Spectrometer	TEMPO (Tropospheric Emissions: Monitoring of Pollution	NASA	NET[2] 2018	O3, NO2, SO2, HCHO, CHOCHO, AOD, AAOD, AI, cloud fraction, cloud top pressure, hourly depending on species and solar zenith angle (SZA)	Geostationary 8 x 4.5 km^2	https://directory.eoportal.org/web/eoportal/satellite-missions/t/tempo
Next Generation Aerosol Polarimetry Sensor (APS), Ocean Ecosystem Spectrometer/ Radiometer (OES)	PACE (Preliminary Aerosol, Cloud, Ecosystem)	NASA	NET[6] 2020	Aerosol and cloud profiles, aerosol optical depth, aerosol subtype	Polar	http://dsm.gsfc.nasa.gov/PACE.html
Cloud radar, High Spectral Resolution Lidar	ACE (Aerosol Clouds and	NASA	NET[6] 2023	Aerosol and cloud profiles, aerosol optical depth, aerosol subtype	Polar	http://dsm.gsfc.nasa.gov/ace/scie nce.html

Instrument	Satellite Platform[4]	Lead Agency	Initiated /Planned	Measurement Parameters[2]	Orbit & Horizontal Resolution	Location of Information and/or Data
(HSRL), Multi-band UV/VIS Spectrometer, Next generation Aerosol Polarimetry Sensor (APS)	Ecosystem Mission)					
CO2 LIDAR	ASCENDS (Active Sensing of CO2 Emissions over Nights, Days, and Seasons)	NASA	NET6 2023	Day/night, all-latitude, all-season CO2 column integrals	Polar	http://cce.nasa.gov/ ascends/index.htm
Event Imaging Spectrometer from GEO, IR Correlation Radiometer, UV/Vis Near IR Wide Imaging Spectrometer	GEO-CAPE (Geostationary Coastal and Air Pollution Events)	NASA	NET[3] 2023	Baseline measurements: O3, NO2, CO, SO2, HCHO, CH4, NH3, CHOCHO, different temporal sampling frequencies; AOD, AAOD, AI, aerosol optical centroid height (AOCH), hourly for solar zenith angle (SZA) <70	Geostationary 4×4 km^2	http://geo-cape.larc. nasa.gov/
SCIAMACHY (Scanning Imaging Absorption Spectrometer for Atmospheric Chartography)	Envisat	ESA	2002	Total column for O3, NO2, BrO, OClO, SO2, HCHO, CHOCHO, aerosols	Polar 60×30 km^2	http://envisat.esa.int/ instruments/sciamachy/
GOME & GOME-2 (Global Ozone Monitoring Experiment)	ERS-2 MetOp-A MetOp-B MetOp-C	EUMET SAT ESA	1995 2006 2012 2016	Total column for O3, NO2, BrO, SO2, HCHO, aerosols	Polar 40×40 km^2	http://earth.esa.int/ers/ gome/ http://www.esa.int/ esaLP/SEMTTEG23 IELPmetop_0.html

(Continued)

Instrument	Satellite Platform[4]	Lead Agency	Initiated /Planned	Measurement Parameters[2]	Orbit & Horizontal Resolution	Location of Information and/or Data
IASI (Infrared Atmospheric Sounding Interferometer)	MetOp-A MetOp-B MetOp-C	ESA	2006 2012 2016	O3, CO, CH4, NH3	Polar 50 x 50 km^2	http://smsc.cnes.fr/IASI/index.htm
TROPOMI (Tropospheric Ozone Monitoring Instrument)	Sentinel-5 Precursor (S5p)	ESA	2015	O3, NO2, HCHO, CHOCHO, SO2, aerosols. aerosol height distribution, CO, and CH4	Polar	http://www.esa.int/esaLP/SEM3ZT4KXMF_LPgmes_0.html
GEMS (Geostationary Environmental Monitoring Sensor)	Geo-KOMPSAT (Geostationary Korea Multi-Purpose Satellite)	KARI NIER-ME	2018	O3, NO2, HCHO, SO2, aerosols	Geostationary	http://database.eohandbook.com/database/missionsummary.aspx?missionID=669
Sentinel 4A (Ultraviolet Visible Near-infrared (UVN) spectrometer) and thermal InfraRed Sounder (IRS)IRS	MTG-S1 (Meteosat Third Generation Sounding Satellite)	EUMETSAT EC, ESA	2019	O3, NO2, HCHO, SO2, aerosols aerosol height distribution CO and CH4	Geostationary	http://www.eumetsat.int/Home/Main/Satellites/MeteosatThirdGeneration/index.htm?l=en

Footnotes:

1. Some instrument systems listed (e.g., UARS/HALOE) are oriented primarily to stratospheric measurements and may have limited application to the troposphere.
2. Note that many of the satellite instruments also have the capability to measure temperature, H_2O, and other parameters.
3. NOAA satellites as early as 1978 have carried AVHRR, and as early as 1985 have carried SBUV/2; as part of a cooperative effort, it is planned to carry AVHRR/3 on MetOp-B in 2012.
4. CALIPSO -- Cloud-Aerosol Lidar & Infrared Pathfinder Satellite Observations.

DMSP -- Defense Meteorological Satellite Program EC -- European Countries.
EOS -- Earth Observing System.
ESA -- European Space Agency.
EUMETSET --European Organisation for the Exploitation of Meteorological Satellites.
GOES -- Geostationary Operational Environmental Satellites KARI -- Korea Aerospace Research Institute.
KMA -- Korean Meteorological Agency.
NASA -- National Aeronautics and Space Administration.
NIER-ME -- National Institute of Environmental Research, Ministry of Environment, Rep. of Korea.
NOAA -- National Oceanic and Atmospheric Administration.
NPOESS -- National Polar-orbiting Operational Environmental Satellite System OCO-2 -- Orbiting Carbon Observatory.
UARS -- Upper Atmosphere Research Satellite.

5. This satellite is part of the A-Train group of satellites. It involves satellites flying in a formation that cross the equator one satellite at a time, a few minutes apart, at around 1:30 pm local time. The A-Train is made up of Aqua, Aura, CALIPSO, and may eventually include OCO-2 and Glory; it also includes CloudSat (2005) – data on the structure of ice and water clouds, and has included PARASOL (2004) – data on the directional characteristics and polarization of light reflected by the Earth and atmosphere, including aerosol optical depth. Together their overlapping science instruments will give a comprehensive picture of Earth weather and climate.
6. “NET” -- No Earlier Than.

APPENDIX I. AIR MONITORING NETWORKS FOR CLIMATE FORCING, TRANSPORT, VERTICAL PROFILE INFORMATION, AND STRATOSPHERIC OZONE

Network	Lead Federal Agency	Number of Sites	Initiated	Measurement Parameters	Location of Information and/or Data
Global Monitoring Division Baseline Observatories					
Mauna Loa	NOAA	1	1957	CO_2, CO, CH_4, SO_2, O_3, SF_6, N_2O, H_2, CFCs, HCFCs, HFC's, Aerosols, Solar Radiation, Surface Meteorology	http://www.esrl.noaa.gov/gmd/o bop/mlo/
Point Barrow	NOAA	1	1973	CO_2, CO, CH_4, SO_2, O_3, SF_6, N_2O, H_2, CFCs, HCFCs, HFC's, Aerosols, Solar Radiation, Surface Meteorology	http://www.esrl.noaa.gov/gmd/o bop/brw/
Samoa	NOAA	1	1974	CO_2, CO, CH_4, SO_2, O_3, SF_6, N_2O, H_2, CFCs, HCFCs, HFC's, Aerosols, Solar Radiation, Surface Meteorology	http://www.esrl.noaa.gov/gmd/o bop/smo/
South Pole	NOAA	1	1957	CO_2, CO, CH_4, SO_2, O_3, SF_6, N_2O, H_2, CFCs, HCFCs, HFC's, Aerosols, Solar Radiation, Surface Meteorology	http://www.esrl.noaa.gov/gmd/o bop/spo/
Trinidad Head	NOAA	1	2002	CO_2, CO, CH_4, SO_2, O_3, SF_6, N_2O, H_2, CFCs, HCFCs, HFC's, Aerosols, Solar Radiation, Surface Meteorology	http://www.esrl.noaa.gov/gmd/o bop/thd/
Summit (Greenland)	NOAA	1	2009	CO_2, CH_4, O_3, SF_6, N_2O, CFCs, Surface Meteorology	http://www.esrl.noaa.gov/gmd/o bop/sum/
Global Monitoring Division					
Observatory Measurements	NOAA	4	1957	See above baseline observatories	http://www.esrl.noaa.gov/gmd/ccgg/insitu.html
Cooperative fixed sites (CCGG)	NOAA	72	1967	CO_2, CH_4, CO, H_2, N_2O, and SF_6, stable isotopes of CO_2 and CH_4	http://www.esrl.noaa.gov/gmd/ccgg/flask.html
Surface Aerosol Measurements	NOAA	29	1976	Absorption, scattering, particle number, and chemical composition	http://www.esrl.noaa.gov/gmd/aero/

Network	Lead Federal Agency	Number of Sites	Initiated	Measurement Parameters	Location of Information and/or Data
Cooperative fixed sites (Ozone)	NOAA	14	1974	Ozone	http://www.esrl.noaa.gov/gmd/o zwv/surfoz/index.html
SURFRAD Measurements (Solar radiation)	NOAA	7	1993	Downwelling short wave, and long wave radiation. Albedo, aerosol optical depth, direct beam and diffuse radiation and UV radiation.	http://www.srrb.noaa.gov/surfrad/
Cooperative fixed sites (Solar Radiation)	NOAA	40	1966	Downwelling short wave, and long wave radiation. Albedo, aerosol optical depth, direct beam and diffuse radiation and UV radiation.	http://www.esrl.noaa.gov/gmd/grad/field.html
Commercial Ships	NOAA	5	1986	CO_2, CH_4, CO, H_2, N_2O, and SF_6, stable isotopes of CO_2 and CH_4	http://www.esrl.noaa.gov/gmd/ccgg/flask.html
University of Washington					
Cheeka Peak Observatory	None	1	1997	O_3, CO, Aerosols, Solar Radiation, Surface Meteorology	http://www.washington.edu/research/field/
Mt. Bachelor Observatory	None	1	2004	O_3, CO, NO/NO_2, Aerosols, Hg, Surface Meteorology	http://research.uwb.edu/jaffegro up/modules/mbo_plot/
Vertical Profile and Other Measurement Networks					
ALE / GAGE / AGAGE Network	NASA	7	1978	CO, CH_4, SF_6, N_2O, H_2, CFCs, HCFCs, HFC's, methyl chloroform, carbon tetrachloride, chloroform, perchloroethylene, halons & others	http://cdiac.ornl.gov/ndps/alegage.html
Tall Tower Measurements (GHG)	NOAA	8	1992	CO_2, CO, CH_4, H_2, CFCs, methyl chloroform, carbon tetrachloride, chloroform, sulfur hexafluoride, perchloroethylene	http://www.esrl.noaa.gov/gmd/ccgg/towers/index.html http://www.nacarbon.org/cgi-nacp/web/investigations/inv_pgp.pl?pgid=171
Aircraft Measurements (GHG)	NOAA	15	1992	CO_2, CH_4, CO, H_2, N_2O, and SF_6, stable isotopes of CO_2 and CH_4	http://www.esrl.noaa.gov/gmd/ccgg/aircraft/

(Continued)

Network	Lead Federal Agency	Number of Sites	Initiated	Measurement Parameters	Location of Information and/or Data
Lidar Aerosol Measurements	NOAA	4	1979	Aerosol light scattering versus altitude	http://www.esrl.noaa.gov/gmd/obop/mlo/programs/gmdlidar/mlo/gmdlidar_mlo.html
NOAA Profiler Network	NOAA	35	1992	Vertical wind and temperature profiles, surface meteorology	http://www.etl.noaa.gov/et7/data/ http://www.profiler.noaa.gov/np n/index.jsp
REALM – Regional East Atmospheric Lidar Mesonet	NOAA	13	2004	Lidar measurements for mixing height and vertical profiling of aerosols, ozone and water vapor	http://alg.umbc.edu/REALM/REALM-LIDAR-Locations.htm
Ozonesonde Network	NOAA	12	1986	Weekly upper air measurements of ozone, temperature, and humidity information from surface to approximately 32 km	http://www.esrl.noaa.gov/gmd/d v/site/site_table.php
SHADOZ Network (Southern Hemisphere ADditional Ozonesondes)	NASA	12	1998	Upper air measurements of ozone, temperature, and humidity	http://croc.gsfc.nasa.gov/shadoz/
Dobson Total Column Ozone	NOAA	19	1962	Ozone	http://www.esrl.noaa.gov/gmd/o zwv/dobson/
Aircraft Measurements (Ozone)	NOAA	9	2004	Ozone	http://www.esrl.noaa.gov/gmd/o zwv/aircraft/aircraft.html
Balloon Water Vapor Measurements	NOAA	3	1980	Water Vapor	http://www.esrl.noaa.gov/gmd/o zwv/wvap/
Cooperative fixed sites (Halocarbons)	NOAA	15	1977	CFC-11 & -12, HCFC-142b, CH2Br2, N2O, CS2, CHBr3, CFC-113, HCFC-22, C6H6, HCFC- 141b, CS2, HFC-134a, HFC-152a, CH3Br, CH3Cl, CH3I, SF6, CH3CCl3,	http://www.esrl.noaa.gov/gmd/h ats/flask/flasks.html

Network	Lead Federal Agency	Number of Sites	Initiated	Measurement Parameters	Location of Information and/or Data
				CCl4, CH2Cl2, CHCl3, C2Cl4, C2HCl, Halon 1211,-1301,- 2402.	
Observatory Measurements (Halocarbons)	NOAA	8	1999	See above baseline observatories	http://www.esrl.noaa.gov/gmd/h ats/insitu/cats/cats_conc.html
Aircraft Measurements (Halocarbons)	NOAA	5	1989	CFC-11 & -12, HCFC-142b, CH2Br2, N2O, CS2, CHBr3, CFC-113, HCFC-22, C6H6, HCFC- 141b, CS2, HFC-134a, HFC-152a, CH3Br, CH3Cl, CH3I, SF6, CH3CCl3, CCl4, CH2Cl2, CHCl3, C2Cl4, C2HCl, Halon 1211,-1301,- 2402.	http://www.esrl.noaa.gov/gmd/ccgg/aircraft/
NEON – The National Ecological Observatory Network	NSF	20 Towers	2012	Site/Flux measurements of CO2, H2O, and meteorological variables, Site measurements of AOD, incoming radiation, dry/wet deposition (nitrogen, carbon, sulfur) Airborne/satellite remote sensing with spectrometer, lidar, digital camera	http://www.neoninc.org/
Ameriflux CO2 exchange network	DOE, NOAA, USDA, NASA	~50 Micrometerological Towers	1996	CO2, meteorological variables	http://ameriflux.ornl.gov/
FluxNet CO2 exchange network	International	~500 Micrometerological Tower Sites	1996	CO2, meteorological variables	http://www.fluxnet.ornl.gov/fluxnet/index.cfm
North American Carbon Program Atmospheric Observing System	Multiple participants	Multiple platforms	2001	CO, CO2, CH4	http://www.nacarbon.org/nacp/

(Continued)

Network	Lead Federal Agency	Number of Sites	Initiated	Measurement Parameters	Location of Information and/or Data
AERONET – AErosol RObotic NETwork MPLNET – Micro-pulse Lidar Network	NASA co-located networks	~500 35	1998 2000	Aerosol spectral optical depths, aerosol size distributions, and precipitable water Aerosols and cloud layer heights	http://aeronet.gsfc.nasa.gov/index.html http://mplnet.gsfc.nasa.gov/
International Aircraft Measurements					
MOZAIC (Measurement of ozone, water vapor, carbon monoxide and nitrogen oxides aboard Airbus in-service aircraft)	None	2500 Airbus international flights/year	1994	O_3, H_2O, CO, NO_x	http://www.iagos.fr/web/rubrique3.html
NOXAR (Measurements of Nitrogen Oxides and Ozone Along Air Routes)	None	500 Swiss Air flights to U.S. and far east	1995 - 1996	O_3, NO, NO_2	http://www.iac.ethz.ch/en/research/chemie/tpeter/Noxar.html
CARIBIC (Civil Aircraft for the Regular Investigation of the atmosphere Based on an Instrument Container)	None	~100 Lufthansa flights	1997	CO, O_3, CO, CH_4, CO_2, N_2O, SF_6, NMHC, Position & Meteorology and Cloud cover.	http://www.caribic-atmospheric.com/
AMATRAS (Atmospheric Measurement by Airliners for Trace Species)	None	262 flights between Japan and Australia	1993	CO_2, CH_4, CO and SF_6	http://www.jal.com/en/press/0000336/img/AMATRAS.pdf

Network	Lead Federal Agency	Number of Sites	Initiated	Measurement Parameters	Location of Information and/or Data
NOAA Research Observing Systems (Systems typically incorporated in intensive field campaigns)					
R/V Ronald H. Brown, Lockheed WP-3D Twin Otter	NOAA	ship aircraft aircraft	N/A	O_3, NO, NO_2, NO_y, VOCs, CO_2, CO, SO_2, HNO_3, NH_3, other reactive pollutants, aerosols, meteorological parameters & upper air, altitude	http://esrl.noaa.gov/csd/projects/2006/data/

Appendix J. Abbreviations and Acronyms

Monitoring-Related Terminology

AOD	Aerosol Optical Depth
AQI	Air Quality Index
AQS	Air Quality System
CAA	Clean Air Act
CFR	Code of Federal Regulations
CMAQ	Community Multiscale Air Quality Model
CTM	Chemical Transport Model
FEM	Federal Equivalent Method
FDDA	Four-Dimensional Data Assimilation
FRM	Federal Reference Method
FTIR	Fourier transform infrared spectroscopy
GHG	Greenhouse Gas
HAPs	Hazardous Air Pollutants
LIDAR	LIght Detection And Ranging
NAAMS	National Ambient Air Monitoring Strategy
NAAQS	National Ambient Air Quality Standards
NATA	National Air Toxics Assessment
NMOC	Non-Methane Organic Carbon
PBL	Planetary Boundary Layer
POP	Persistent Organic Pollutant
RASS	Radio-Acoustic Sounding System
SIP	State Implementation Plan
VOC	Volatile Organic Compound

Government Agencies and Sponsored Organizations

AQRS	Air Quality Research Subcommittee (CENRS)
CENRS	Committee on Environment, Natural Resources, and Sustainability
CEQ	Council on Environmental Quality
DHHS	Department of Health and Human Services
DHS	Department of Homeland Security
DOE	Department of Energy

DOI	Department of the Interior
EPA	Environmental Protection Agency
ESA	European Space Agency
EUMETSAT	European Organisation for the Exploitation of Meteorological Satellites
GEO	Group on Earth Observations
GMES	Global Monitoring for Environment and Security
HEI	Health Effects Institute
IGAC	International Global Atmospheric Chemistry Project
NACP	North American Carbon Program
NARSTO	North American Research Strategy for Tropospheric Ozone
NAS	National Academy of Sciences
NASA	National Aeronautics and Space Administration
NCAR	National Center for Atmospheric Research
NOAA	National Oceanic and Atmospheric Administration
NPS	National Park Service
NRC	National Research Council
NSF	National Science Foundation
NSTC	National Science and Technology Council
NWS	National Weather Service
OFCM	Office for the Federal Coordinator for Meteorology
OMB	Office of Management and Budget
OSTP	Office of Science & Technology Policy
USGEO	United States Group on Earth Observations
USDA	United States Department of Agriculture
USFS	United States Forest Service
USGS	United States Geological Survey
WMO	World Meteorological Organization

Monitoring Networks

AERONET	Aerosol Robotic NETwork
AGAGE	Advanced Global Atmospheric Gases Experiment
AIRMoN	Atmospheric Integrated Research Monitoring Network
ALE	Atmospheric Lifetime Experiment
AMNET	Atmospheric Mercury Network (NADP)

AMoN	Ammonia Monitoring Network (NADP)
ASOS	Automated Surface Observing System (NOAA)
BSRN	Baseline Surface Radiation Network
CARIBIC	Civil Aircraft for Regular Investigation of the atmosphere Based on an Instrument Container
CASTNET	Clean Air Status and Trends Network
CSN	Chemical Speciation Network
GAGE	Global Atmospheric Gases Experiment
GALION	GAW Atmospheric Lidar Observation Network
GAW	Global Atmospheric Watch
GEMS	Global & regional Earth-system (Atmosphere) Monitoring using Satellite and in-situ data
GEOSS	Global Earth Observation System of Systems
IADN	Integrated Atmospheric Deposition Network
IAGOS	In-Service Aircraft for the Global Observing System
IGACO	Integrated Global Atmospheric Chemistry Observations
IMPROVE	Interagency Monitoring of Protected Visual Environments
MACC	Monitoring Atmospheric Composition and Climate
MADIS	Meteorological Data Ingest System
MDN	Mercury Deposition Network
MOZAIC	Measurement of Ozone, water vapor, carbon monoxide and nitrogen oxides aboard inservice Airbus aircraft
MPLNET	Micro Pulse Lidar Network
NADP	National Atmospheric Deposition Program (NADP)
NATTS	National Air Toxics Trends Stations
NCore	National Core Network
NDACC	Network for the Detection of Atmospheric Composition Change
NEUBREW	NOAA-EPA Brewer Spectrophotometer UV and Ozone Network
NPN	NOAA Profiler Network
NTN	National Trends Network (NADP)
PAMS	Photochemical Assessment Monitoring Stations
SEARCH	South Eastern Aerosol Research and Characterization Study
REALM	Regional East Atmospheric Lidar Mesonet
SIRD	Supersites Integrated Relational Database

SLAMS	State and Local Air Monitoring System
SOS	Southern Oxidants Study
SURFRAD	SURFace RADiation budget observing network
TAMDAR	Tropospheric Airborne Meteorological Data Reporting
TOLNet	Tropospheric Ozone Lidar Network
VIEWS	Visibility Information Exchange Web System
UATMP	Urban Air Toxics Monitoring Program

Intensive Field Campaigns

ARCTAS	Arctic Research of the Composition of the Troposphere from Aircraft and Satellites
AUSPEX	Atmospheric Utility Signatures, Predictions and Experiments
DISCOVER	AQ – Deriving Information on Surface Conditions from Column and Vertically Resolved Observations Relevant to Air Quality
EMEFS	Eulerian model evaluation field study
HTAP	Hemispheric Transport of Air Pollution
ICARTT	International Consortium for Atmospheric Research on Transport and Transformation
INTEX-NA	Intercontinental Chemical Transport Experiment North America
INTEX-B	Intercontinental Chemical Transport Experiment Phase B
ITCT	Intercontinental Transport and Chemical Transformation
MILAGRO	Megacity Initiative: Local and Global Research Observations
NARE	North American Regional Experiment
NEAQS	ITCT 2004 – New England Air Quality Study – Intercontinental Transport and Chemical Transformation
POLARCAT	Polar Study using Aircraft, Remote Sensing, Surface Measurements and Models, of Climate Chemistry, Aerosols, and Transport
RAPS	Regional Air Pollution Study
SJVAQS	San Joaquin Valley Air Quality Study
TexAQS	Texas Air Quality Study
TRACE-P	Transport and Chemical Evolution over the Pacific

Satellite-Oriented Programs, Systems & Terms

ACE	Aerosol-Cloud-Ecosystems mission
EOS	Earth Orbiting System
CALIPSO	Cloud-Aerosol Lidar and Infrared Pathfinder Satellite Observation satellite
GEO	Geostationary satellite platform
GEO-CAPE	Geostationary Coastal and Air Pollution Events
GACM	Global Atmospheric Composition Mission
GOES	Geostationary Operational Environmental Satellites
GOME	Global Ozone Monitoring Experiment
GOSAT	Greenhouse gases Observing SATellite
JPSS	Joint Polar Satellite System
LEO	Low Earth Orbit
MODIS	Moderate Resolution Imaging Spectroradiometer
MOPITT	Measurements of Pollution in the Troposphere
NESDIS	National Environmental Satellite, Data, and Information Service (NOAA)
NPOESS	National Polar-orbiting Operational Environmental Satellite System
NPP	NPOESS Preparatory Project
OCO	Orbiting Carbon Observatory
OMI	Ozone Monitoring Instrument
PARASOL	Polarization and Anisotropy of Reflectances for Atmospheric Sciences coupled with Observations from a Lidar
POES	Polar Operational Environmental Satellite
PROMOTE	PROtocol MOniToring for the GMES Service Element: Atmosphere
SCIAMACHY	SCanning Imaging Absorption SpectroMeter for Atmospheric CHartographY
TEMPO	Tropospheric Emissions: Monitoring of Pollution
TOMS	Total Ozone Mapping Spectrometer
TOR	Tropospheric Ozone Residual

REFERENCES

Al-Saadi, J., Szykman, J., Pierce, R. B., Kittaka, C., Neil, D., Chu, D. A., Remer, L., Gumley, L., Prins, E., Weinstock, L., MacDonald, C., Wayland, R., Dimmick, F. & Fishman, J. (2005). Improving national air quality forecasts with satellite aerosol observations. *Bull. Am. Met. Soc.*, *86*, 1249-1261.

Committee on Earth Observation Satellites. (2011). A geostationary satellite constellation for observing global air quality: An international path forward. CEOS Atmospheric Composition Constellation, Draft ACC White Paper v4, 41 (available at http://www.ceos.org/index.php?option=com_content&view=article&id=64:2008-07-25-14-45-59&catid=54:acc-documents&Itemid=95)

Cooper, O. R., Parrish, D. D., Stohl, A., Trainer, M., Nedelec, P., Thouret, V., Cammas, J. P., Oltmans, S. J., Johnson, B. J., Tarasick, D., Leblanc, T., McDermid, I. S., Jaffe, D., Gao, R., Stith, J., Ryerson, T., Aikin, K., Campos, T., Weinheimer, A. & Avery, M. A. (2010). Increasing springtime ozone mixing ratios in the free troposphere over western North America. *Nature*, *463*, 344-348.

Duncan, B., Yoshida, Y., Olson, J., Sillman, S., Martin, R., Lamsal, L., Hu, Y., Pickering, K., Retscher, C., Allen, D. & Crawford, J. (2010). Application of OMI observations to a space-based indicator of NOx and VOC controls on surface ozone formation. *Atmos. Environ.*, *44*, 2213-2223, doi:10.1016/j.atmosenv.2010.03.010.

Engel-Cox, J. A., Holloman, C. H., Coutant, B. W., Hoff, R. M. (2004). Qualitative and quantitative evaluation of MODIS satellite sensor data for regional and urban scale air quality. *Atmos. Env.*, *38*, 2495-2509.

Environmental Protection Agency. (1997). *National Ambient Air Quality Standards for Particulate Matter*. CFR, Part 50, Title 40.

Environmental Protection Agency. (2006). *Revisions to Ambient Air Monitoring Regulations*. CFR Parts 53 and 58, 71 FR 61236, October 17, 2006.

Environmental Protection Agency. (2008). Ambient Air Monitoring Strategy for State, Local, and Tribal Air Agencies. Office of Air Quality Planning and Standards, Research Triangle Park, NC. (available at http://www.epa.gov/ttn/amtic/monstratdoc.html)

Environmental Protection Agency. (2010). *Primary National Ambient Air Quality Standards for Nitrogen Dioxide*. 40 CFR Parts 50 and 58, 71 FRL 9107-9, February 9, 2010. (available at http://www.epa.gov/ttn/naaqs/standards/nox/fr/20100209.pdf)

Fehsenfeld, Trainer, M., Parrish, D. D., Volz-Thomas, A. & Penkett, S. (1996). North Atlantic Regional Experiment 1993 summer intensive: Foreword. *J. Geophys. Res.*, *101*, 28, 869-28,875.

Fehsenfeld, F. C., Ancellet, G., Bates, T. S., Goldstein, A. H., Hardesty, R. M., Honrath, R., Law, K. S., Lewis, A. C., Leaitch, R., McKeen, S., Meagher, J., Parrish, D. D., Pszenny, A. A. P., Russell, P. B., Schlager, H., Seinfeld, J., Talbot, R. & Zbinden, R. (2006). International Consortium for Atmospheric Research on Transport and Transformation (ICARTT): North America to Europe – Overview of the 2004 summer field study. *J. Geophys. Res.*, *111*, D23S01, doi:10.1029/2006JD007829.

Fioletov, V. E., McLinden, C. A., Krotkov, N., Moran, M. D. & Yang, K. (2011). Estimation of SO_2 emissions using OMI retrievals. *Geophysical Research Letters*, Vol. *38*, L21811, doi:10.1029/2011GL049402.

Fishman, J. & Crutzen, P. J. (1978). The Origin of Ozone in the Troposphere. *Nature*, *274*, 855- 858.

Fishman, J., et al., (2008). Remote Sensing of Tropospheric Pollution from Space. *Bull. Am. Met. Soc.*, *89*(6), 805-821.

Fishman et al. (2012). The United States' Next Generation of Atmospheric Composition and Coastal Ecosystem Measurements: NASA's Geostationary Coastal and Air Pollution Events (GEO- CAPE) Mission. Bull. Am. Met. Soc. doi: 10.1175/BAMS-D-11-00201.1 http://journals.ametsoc.org/doi/pdf/10.1175/BAMS-D-11-00201.1

Geller, M. D. & Solomon, P. A. (2006). Special Issue of Aerosol Science and Technology for Particulate Matter Supersites Program and Related Studies. *Aerosol Sci. and Tech.*, *40*, 10,735- 736.

Ingmann, P., Veihelmann, B., Langen, J., Lamarre, D., Stark, H., Bazalgette Courrèges-Lacoste, G. (2012). Requirements for the GMES Atmosphere Service and ESA's implementation concept: Sentinels-4/-5 and -5p. *Remote Sensing of Environment*, Volume *120*, 15 May 2012, Pages 58-69, ISSN 0034-4257, 10.1016/j.rse.2012.01.023. (http://www.sciencedirect.com/science/article/pii/S0034425712000673)

Keating, T. & Zuber, A. (2007). Hemispheric Transport of Air Pollution 2007, *Interim Report prepared by the Task Force on Hemispheric Air Pollution.* United Nations Publication ISSN 1014- 4625.

Kondragunta, S., Lee, P., McQueen, J., Kittaka, C., Prados, A. I., Ciren, P., Laszlo, I., Pierce, R. B., Hoff, R. & Szyman, J. (2008). Air Quality Forecast Verification Using Satellite Data. AMS, DOI: 10.1175/ 2007JAMC1392.1

Lahoz, W. A., Peuch, V. -H., Orphal, J., Attié, J. -L., Chance, K., Liu, X., Edwards, D., Elbern, H., Flaud, J. -M., Claeyma n, M. & El Amraoui, L. (2012). Monitoring Air Quality From Space: The Case for the Geostationary Platform. *Bull. Am. Met. Soc*, DOI:10.1175/BAMS-D-11-00045.1

Lamsal, L. N., Martin, R. V., Padmanabhan, A., van Donkelaar, A., Zhang, Q., Sirois, C. E., Chance, K., Kurosu, T. P. & Newchurch, M. J. (2011). Application of satellite observations for timely updates to global anthropogenic NOx emissions inventories. *Geophysical Research Letters*, Vol. *38*, L05810, doi:10.1029/2010GL046476.

Liu, Y., Park, R. J., Jacob, D. J., Li, Q. B., Kilaru, V. & Sarnat, J. A. (2004). Mapping annual mean ground-level $PM_{2.5}$ concentrations using Multiangle Imaging Spectroradiometer aerosol optical thickness over the contiguous United States. *Journal of Geophysical Research-Atmospheres*, 2004, *109*, (D22).

Liu, X., Chance, K., Sioris, C. E., Kurosu, T. P., Spurr, R. J. D., Martin, R. V., Fu, T., Logan, J. A., Jacob, D. J., Palmer, P. I., Newchurch, M. J., Megretskaia, I. A. & Chatfield, R. B. (2006). First directly retrieved global distribution of tropospheric column ozone from GOME: Comparison with the GEOS-CHEM model. *J. Geophys. Res.*, *111*, D02308, doi:10.1029/2005JD006564.

Liu, X., Bhartia, P. K., Chance, K., Spurr, R. J. D. & Kurosu, T. P. (2009). Ozone profile retrievals from the Ozone Monitoring Instrument. *Atmos. Chem. Phys. Discuss.*, *9*, 22693-22738.

Lu, Z., Streets, D. G., Zhang, Q., Wang, S., Carmichael, G. R., Cheng, Y. F., Wei, C., Chin, M., Diehl, T. & Tan, Q. (2010). Sulfur dioxide emissions in China and sulfur trends in East Asia since 2000. *Atmos. Chem. Phys.*, *10*, 6311-6331.

Martin, R. V., Fiore, A. & van Donkelaar, A. (2004). Space-based diagnosis of surface ozone sensitivity to anthropogenic emissions. *Geophys. Res. Lett.*, *31*, L06120, doi:10.1029/2004GL019416.

Martin, R. V., Sioris, C. E., Chance, K., Ryerson, T. B., Bertram, T. H., Wooldridge, P. J., Cohen, R. C., Neuman, J. A., Swanson, A. & Flocke, F. M. (2006). Evaluation of space-based constraints on global nitrogen oxide emissions with regional aircraft measurements over and downwind of eastern North America. *J. Geophys. Res.*, *111*, D15308, doi:10.1029/2005JD006680.

Martin, R. V. (2008). Satellite remote sensing of surface air quality. *Atmos. Environ.*, *42*, 7823- 7843.

Mathur, R. (2008). Estimating the impact of the 2004 Alaskan forest fires on episodic particulate matter pollution over the eastern United States through assimilation of satellite-derived aerosol optical depths in a regional air quality model. *J. Geophys. Res.*, doi:10.1029/2007JD009767.

McLinden, C. A., Fioletov, V., Boersma, K. F., Krotkov, N., Sioris, C. E., Veefkend, J. P. & Yang, K. (2012). Air quality over the Canadian oil sands: A first assessment using satellite observations. *Geophysical Research Letters*, Vol. *39*, L04804, doi:10.1029/2011GL050273.

Millet, D. B., Jacob, D. J., Boersma, K. F., Fu, T., Kurosu, T. P., Chance, K., Heald, C. L. & Guenther, A. (2008). Spatial distribution of isoprene emissions from North America derived from formaldehyde column measurements by the OMI satellite sensor. *J. Geophys. Res.*, *113*, D02307, doi:10.1029/2007JD008950.

National Research Council (NRC). (1991). *Rethinking the Ozone Problem in Urban and Regional Air Pollution.* National Academy of Sciences Press, Washington, DC.

National Research Council (NRC). (2004). *Air Quality Management in the United States.* The National Academies Press, Washington, DC.

National Research Council (NRC). (2007). Earth Science and Applications from Space: *National Imperatives for the Next Decade and Beyond.* The National Academies Press, Washington, DC.

National Research Council (NRC). (2009a). *Global Sources of Local Pollution: An Assessment of Long-Range Transport of Key Air Pollutants to and from the United States.* The National Academies Press, Washington, DC.

National Research Council (NRC). (2009b). *Observing Weather and Climate from the Ground Up: A Nationwide Network of Networks.* The National Academies Press, Washington, DC.

Neil, D. O., Kondragunta, S., Osterman, G., Pickering, K., Pinder, R. W., Prados, A. & Szykman, J. (2009). *Satellite Observations for Detecting and Tracking Changes in Atmospheric Composition*. EM, October 2009, 13-15.

Pandis, S., Solomon, P. A. & Scheffe, R. (2005). Preface to special section on Particulate Matter Supersites. *J. Geophys. Res.*, doi:10.1029/2005 JD005983.

Penkett, S. A., Volz-Thomas, A., Parrish, D. D., Honrath, R. E. & Fehsenfeld, F.C. (1998). North Atlantic Regional Experiment (NARE II): Preface. *J. Geophys. Res.*, *103*, 13, 353-13,355.

Peters, A., Dockery, D. W., Muller, J. E. & Middleton, M. A. (2001). Increased particulate air pollution and the triggering of myocardial infarction. *Circulation*, *103*, 2810-2815.

Pfister, G., Hess, P. G., Emmons, L. K., Lamarque, J.-F., Wiedinmyer, C., Edwards, D. P., Pétron, G., Gille, J. C. & Sachse, G. W. (2005): Quantifying CO emissions from the 2004 Alaskan wildfires using MOPITT CO data. *Geophys. Res. Lett.*, *32*, L11809, doi:10.1029/ 2005GL022995.

Pierce, R. B., Al-Saadi, J., Kittaka, C., Schaack, T., Lenzen, A., Bowman, K., Szykman, J., Soja, A., Ryerson, T., Thompson, A. M., Bhartia, P. & Morris, G. A. (2009): Impacts of background ozone production on Houston and Dallas, Texas air quality during the Second Texas Air Quality Study field mission. *Journal of Geophysical Research – Atmospheres*, *114*. D00F09 (2009): doi:10.1029/2008JD011337.

Pinder, R. W., Walker, J. T., Bash, J. O., Cady-Pereira, K. E., Henze, D. K., Luo, M., Osterman, G. B. & Shephard, M. W. (2011): Quantifying spatial and seasonal variability in atmospheric ammonia with in situ and space-based observations. *Geophysical Research Letters*, Vol. *38*, L04802, doi: 10.1029/2010GL046146.

Pope, C. A., Ezzati, M. & Dockery, D. W. (2009). Fine-particulate air pollution and life expectancy in the United States. *N Engl J Med*, *360*, 376-386.

Scheffe, R. D., Solomon, P. A., Husar, R., Hanley, T., Schmidt, M., Koerber, M., Gilroy, M., Hemby, J., Watkins, N., Papp, M., Rice, J., Tikvart, J. & Valentinetti, R. (2009). The National Ambient Air Monitoring Strategy: Rethinking the Role of National Networks. *J. Air & Waste Manage. Assoc.*, *59*, 1-12.

Soja, A. J., Al-Saadi, J., Giglio, L., Randall, D., Kittaka, C., Pouliot, G., Kordzi, J. J., Raffuse, S., Pace, T. G., Pierce, T. E., Moore, T., Roy, B., Pierce, R. B. & Szykman, J. J. (2009). Assessing satellite-based fire data for use in the National Emissions Inventory. *Journal of Allied Remote Sensing*, Vol. *3*, 031504.

Solomon, P., Cowling, E., Hidy, G. & Furiness, C. (2000). Comparison of scientific findings from major ozone field studies in North America and Europe. *Atmospheric Environment*, Vol: *34*, Issue: 12-14, Pages: 1885-1920.

Solomon, P. A., Hopke, P. K., Froines, J. & Scheffe, R. (2008). Key Scientific and Policy- and Health- Relevant Findings from the U.S. EPA's Particulate Matter Supersites Program and Related Studies: An Integration and Synthesis of Results. J. *Air & Waste Manage. Assoc.*, *58*, S-1 – S-92.

Stutz, J., Alicke, B., Ackermann, R., Geyer, A., Wang, S. H., White, A. B., Williams, E. J., Spicer, C. W. & Fast, J. D. (2004). Relative humidity dependence of HONO chemistry in urban areas. *J. Geophys. Res.*, *109*, doi:10.1029/2003JD004135.

Veefkind, J. P., Aben, I., McMullan, K., Förster, H., de Vries, J., Otter, G., Claas, J., Eskes, H. J., de Haan, J. F., Kleipool, Q., van Weele, M., Hasekamp, O., Hoogeveen, R., Landgraf, J., Snel, R., Tol, P., Ingmann, P., Voors, R., Kruizinga, B., Vink, R., Visser, H., Levelt, P. F. (2012). TROPOMI on the ESA Sentinel-5 Precursor: A GMES mission for global observations of the atmospheric composition for climate, air quality and ozone layer applications. *Remote Sensing of Environment*, Vol. *120*, 15 May 2012, Pages 70-83, ISSN 0034-4257, 10.1016/j.rse.2011.09.027.

Vijayaraghavan, K., Snell, H. & Seigneur, C. (2007). *Feasibility of Using Satellite Data in Air Quality Modeling*. CRC Report No. A-61, Coordinating Research Council, Inc., Alpharetta, GA, June 2007.

Wahlin, P., Palmgren, F., Van Dingenen R. (2001). Experimental studies of ultrafine particles in streets and the relationship to traffic. *Atmos. Environ.*, *35*, S63–S69.

Wang, S., Streets, D. G., Zhang, Q., He, K., Chen, D., Kang, S., Lu, Z. & Wang, Y. (2010). Satellite detection and model verification of NOx emissions from power plants in Northern China. *Environ. Res. Lett. 5*, 044007.

West, J. J., Fiore, A. M., Naik, V., Horowitz, L. W., Schwarzkopf, M. D. & Mauzerall, D. L. (2007). Ozone air quality and radiative forcing consequences of changes in ozone precursor emissions. *Geophysical Research Letters*, Vol. *34*, L06806, doi:10.1029/2006GL029173.

Zhou, X., Civerolo, K., Dai, H., Huang, G., Schwab, J. & Demerjian, K. (2002). Summertime nitrous acid chemistry in the atmospheric boundary layer at a rural site in New York State. *J. Geophys. Res*, *107*, 10.1029/2001JD00153971.

Zhu, Y., Hinds, W.C., Kim, S. & Sioutas, C. (2002). Concentration and Size Distribution of Ultrafine Particles near a Major Highway. *J. Air & Waste Manage. Assoc.*, *52*, 1032-10.

End Note

[1] National Aeronautics and Space Administration Authorization Act of (2010 (Public Law 111–267), § 703 "Interagency Collaboration Implementation Approach."

In: Air Quality Observation in the US
Editor: Malcolm Parisi
ISBN: 978-1-63117-154-3

Chapter 2

OZONE AIR QUALITY STANDARDS: EPA'S 2013 REVISION*

James E. McCarthy

SUMMARY

The Environmental Protection Agency (EPA) is nearing the end of a statutorily required review of the National Ambient Air Quality Standards (NAAQS) for ozone. EPA projects proposal of a revised standard by December 2013, with promulgation late in 2014.

NAAQS are standards for outdoor (ambient) air that are intended to protect public health and welfare from harmful concentrations of pollution. If the EPA Administrator changes the standard to a lower level, he or she would be concluding that protecting public health and welfare requires lower concentrations of ozone pollution than it previously judged to be safe. In high enough concentrations, ozone has been found to aggravate heart and lung diseases and may contribute to premature death.

The ozone standards affect a large percentage of the population: as of December 2012, 123 million people (40% of the U.S. population) lived in areas classified nonattainment for the primary (health-based) ozone NAAQS. If the standards are further strengthened, more areas would likely be affected, and those already considered nonattainment and

* This is an edited, reformatted and augmented version of a Congressional Research Service publication, CRS Report for Congress R43092, prepared for Members and Committees of Congress, from www.crs.gov, dated May 30, 2013.

sources that contribute to the nonattainment might have to impose more stringent emission controls.

EPA last revised the ozone standard in March 2008, but the level chosen at that time remains subject to controversy. A 23-member panel of EPA science advisers, chosen from outside the agency, unanimously recommended a more stringent range of standards than the Administrator chose. With the change of Administrations in 2009, the agency agreed to reconsider the 2008 standard, but the process was short-circuited by a Presidential decision to await conclusion of the next regular review—the review now nearing completion.

The agency begins a NAAQS review by compiling an Integrated Science Assessment, which summarizes the science surrounding the standards. The current assessment, released in February 2013, finds that the evidence associating ozone exposure with morbidity and mortality has strengthened since the 2008 review was completed. This would appear to support more stringent standards.

Proposed standards might raise a number of issues, including whether their expected benefits justify their costs. This is a perennial issue raised by stakeholders when EPA considers revising the NAAQS, although, as the Clean Air Act is currently written, the agency is prohibited from weighing costs against benefits in setting these standards. The statute simply states that the Administrator is to set the primary standard at a level requisite to protect the public health, allowing an adequate margin of safety. A unanimous Supreme Court has found that the absence of language mentioning cost means that costs are not to be considered in setting these standards. Many in Congress would like to change this to require such a cost or cost-benefit consideration: in the 112th Congress, House-passed legislation would have done so.

A related question is how nonattainment areas would lower emissions sufficiently to comply with a more stringent standard. Current federal standards for cars, trucks, nonroad vehicles and engines, power plants, and other stationary pollution sources are not strong enough to bring many areas into attainment, thus requiring local pollution control measures in those cases. To attain more stringent standards, there might be increased pressure to strengthen federal controls on the sources that contribute to ozone pollution; Congress might also revisit the perennial issue of how to control sources that contribute to pollution downwind, in other states.

There are also monitoring issues: at present, only 675 of the nation's 3,000 counties have ozone monitors in place. EPA, the states, and Congress may consider whether the current monitoring network is adequate to detect violations of a more stringent standard; both EPA and state monitoring budgets are constrained, however, raising questions as to how any additional monitoring requirement would be funded.

This report discusses the standard-setting process, the specifics of the current and most recent reviews, and issues that may be raised as EPA brings the current review to completion.

INTRODUCTION

The Environmental Protection Agency (EPA) expects to complete a statutorily required review of the National Ambient Air Quality Standards (NAAQS) for ozone by the end of 2014. The agency has already completed the science assessment that will serve as the review's foundation,[1] and has said it will propose any revision to the standards by late 2013. A comment period, public hearings, and interagency review will follow, before the standards are finalized, presumably by the end of the following year.

Ground-level ozone (often referred to as "smog") is associated with serious health effects when present in high enough concentrations. These health effects include aggravated asthma, chronic bronchitis, and heart attacks, and in some cases premature death. In its 2010 proposal to reconsider the ozone standard, EPA said that reducing concentrations to 60 parts per billion (versus the then-current primary standard of 84 parts per billion) would avoid 4,000 to 12,000 premature deaths annually.[2]

High ozone concentrations also affect the growth of plants, causing damage to both forests and field crops. In 2006, the U.S. Forest Service examined 380 monitoring sites in the Mid-Atlantic and Southeastern states and found visible injury to forest plants from ozone at 121 (32%) of them. At 20 of the sites, the damage was described as "severe."[3] In addition, EPA found that "several economically important crop species are sensitive to ozone levels typical of those found in the United States," and estimated that crop losses could be reduced by $400 million to $620 million annually by implementation of a more stringent ozone standard.[4]

While EPA's analysis has found that there would be substantial benefits to reducing ozone, the agency also has concluded that there could be substantial costs. At the time it promulgated its 2008 revision of the standard, EPA projected the cost of attaining the standard at $7.6 billion to $8.8 billion annually.[5] Tightening the standard further might add additional billions to that cost.

The ozone NAAQS are among EPA's most far-reaching standards. At the current level, set in March 2008, 123 million people live in areas that have not attained the standards (see *Figure 1*).[6] These 46 areas (referred to as "nonattainment areas") generally coincide with metropolitan areas, but may be larger or smaller.

Source: U.S. EPA Green Book, http://www.epa.gov/oaqps001/greenbk/map8hr_2008.html. Map shows areas designated nonattainment by EPA as of December 14, 2012.

Notes: Nonattainment designations were based on 2008-2010 monitoring data in most cases. As shown in Figure 2, some areas that attain the 2008 standard are still designated nonattainment for the less stringent 1997 standard. These areas may have monitoring data that demonstrate attainment with the 1997 (and the 2008) standards, but they must have an approved maintenance plan in place before they will be removed from the list of nonattainment areas.

Figure 1. Ozone Nonattainment Areas (2008 Standard).

Nonattainment areas are required to develop plans (State Implementation Plans, SIPs) that demonstrate how emissions will be reduced sufficiently to attain the standards. SIPs are submitted to EPA for approval. Depending on the severity of pollution, the areas have from 3 to 20 years to reach attainment.

In addition to the 46 nonattainment areas for the 2008 standard, 41 areas with 118 million people (mostly the same areas) are still considered nonattainment for the less stringent 1997 ozone NAAQS, and 72 areas with 48 million people are classified as "maintenance areas" for that standard.[7] (See *Figure 2.)* Some of these nonattainment areas may have monitoring data that demonstrate attainment with the 1997 standards; but they must have an EPA-approved maintenance plan in place before they will be removed from the nonattainment list and placed in the "maintenance" category. An approved maintenance plan includes enforceable measures demonstrating how an area will control emissions sufficiently to maintain compliance with the NAAQS.

EPA has several carrots or sticks available to get areas to comply with a NAAQS. The most frequently mentioned of these is the potential for highway-fund sanctions: failure to submit or implement a SIP adequate to attain or maintain compliance with the NAAQS can lead to the temporary suspension of federal highway funds for non-safety-related projects. Ultimately, EPA can impose a federal implementation or maintenance plan (a FIP) in an area that does not have an approved SIP. Imposition of sanctions or FIPs is relatively rare, however: generally the states avoid sanctions by submitting plans that require sufficient emission reductions to be deemed adequate by EPA.

The potential economic, health, and environmental impacts of a change in the ozone NAAQS have led to great interest in EPA's ongoing review of the standards. To assist Members and staff in evaluating EPA's review, this report provides background on NAAQS, the process used to establish them, the current ozone standards, the remaining controversy over the most recent revision, and issues that may be raised as EPA brings the current review to completion.

What Are NAAQS?

NAAQS are standards that apply to ambient (outdoor) air. Section 109 of the Clean Air Act directs EPA to set both primary NAAQS, which are standards, "the attainment and maintenance of which in the judgment of the [EPA] Administrator ... are requisite to protect the public health," with "an adequate margin of safety," and secondary NAAQS, which are standards necessary to protect public welfare, a broad term that includes damage to crops, vegetation, property, building materials, climate, etc.[8]

The pollutants to which NAAQS apply are generally referred to as "criteria" pollutants. The act defines them as pollutants that "endanger public health or welfare," and whose presence in ambient air "results from numerous or diverse mobile or stationary sources."[9] Six pollutants are currently identified as criteria pollutants: ozone, particulates, carbon monoxide, sulfur dioxide, nitrogen oxides, and lead. The EPA Administrator can add to this list if he or she determines that additional pollutants meet the act's criteria, or delete pollutants from the list if they no longer meet the criteria. There have been no changes to the list, however, since the late 1970s.

Source: U.S. EPA Green Book, http://www.epa.gov/oaqps001/greenbk/map8hrnm.html. Map shows areas designated nonattainment or maintenance as of 12/14/2012.

Figure 2. Ozone Nonattainment and Maintenance Areas (1997 Standard).

NAAQS are at the core of the Clean Air Act, even though they do not directly regulate emissions. In essence, they are standards that define what EPA considers to be clean air. Once a NAAQS has been set, the agency, using monitoring data and other information submitted by the states, identifies areas that exceed the standards and must reduce pollutant concentrations. This designation process is often delayed by litigation over the standards, by EPA's agreement to reconsider aspects of them, or by consultations with the states over the specifics of the areas to be designated. Designation of nonattainment areas for the 1997 NAAQS, for example, took seven years. Designations under the 2008 standards took four years.

After nonattainment areas are designated, state and local governments have up to three years to produce State Implementation Plans which outline the measures that will reduce emission levels and attain the standards. Actual attainment of the standards can stretch over a 3-year to 20-year period, depending on the severity of the area's pollution. Ozone nonattainment areas are designated as Marginal, Moderate, Serious, Severe, or Extreme, depending on the level of pollution. Each of these classifications comes with required pollution control measures: the more severe the pollution, the more stringent are the required controls, and the longer the area is allowed before it must demonstrate attainment.[10]

Thus, establishment or revision of a NAAQS is not an event that requires immediate compliance with an air quality standard; rather, it sets in motion a long and complicated implementation process. That process may ultimately have far-reaching impacts for public health, for sources of pollution in numerous economic sectors, and for states and local governments.

In addition to state and local actions to address ambient concentrations of NAAQS pollutants, EPA itself acts to control emissions and concentrations of criteria pollutants, through national standards for products that contribute to ozone pollution (particularly mobile sources, such as automobiles) and emission standards for new stationary sources (such as power plants). These standards lead to emission reductions that states can factor into their implementation plans, reducing the need for local air pollution control measures.

REVIEWING THE OZONE NAAQS

Section 109(d) of the Clean Air Act requires the agency to review each NAAQS every five years. That schedule is rarely met, but it often triggers lawsuits that force the agency to undertake a review. In the case of ozone, the American Lung Association filed suit over EPA's failure to complete a review in 2003, for example, and a consent decree established a schedule under which the Bush Administration's EPA promulgated revisions to the NAAQS in March 2008.[11] After the standard was promulgated, numerous parties petitioned the courts for a review of the standard.

With the change of Administrations, EPA itself concluded that it had concerns whether the 2008 standards "satisfy the requirements of the Clean Air Act."[12] Thus, in September 2009, the agency suspended implementation of the 2008 standards and set a schedule for their reconsideration under which a January 2010 proposal was released.

That reconsideration was terminated in September 2011, when the President sent a nearly final revision of the NAAQS back to EPA and announced that the agency would instead complete its next scheduled five-year review before promulgating any change to the ozone standard.[13] It is that review that is now nearing completion.

The NAAQS Review Process

Reviewing an existing NAAQS is generally a long process. To begin the process, EPA scientists compile the scientific literature published since the last NAAQS revision, and summarize it in a report known as a Criteria Document or Integrated Science Assessment (ISA). The ISA for ozone, completed in February 2013,[14] reviewed 2,300 scientific studies. Ozone ISAs cover topics as wide-ranging as the physics and chemistry of ozone in the atmosphere; environmental concentrations, patterns, and exposure; dosimetry and animal-to-human extrapolation; toxicology; interactions with co-occurring pollutants; controlled human exposure studies; epidemiology; effects on vegetation and ecosystems; effects on UVB exposures and climate; and effects on manmade materials.

Following completion of the ISA, EPA prepares a Risk and Exposure Assessment to identify exposure pathways, at-risk populations, and health endpoints.

A final document prepared by EPA staff, the Staff Paper or Policy Assessment, summarizes the information compiled in the ISA and Risk Assessment and provides the Administrator with options regarding the indicators, averaging times, statistical form, and numerical level (concentration) of the NAAQS.

To ensure that NAAQS reviews meet the highest scientific standards, the 1977 amendments to the Clean Air Act required the Administrator to appoint an independent Clean Air Scientific Advisory Committee (CASAC). CASAC has seven members, largely from academia and from private research institutions. In conducting NAAQS reviews, their expertise is supplemented by panels of the nation's leading experts on the health and environmental effects of the specific pollutants that are under review. These panels can be rather large. The panel for the 2008 ozone review, for example, had 23 members. CASAC and the public make suggestions regarding the membership of the panels on specific pollutants, with the final selections made by EPA. The panels evaluate the agency's work during NAAQS-setting and NAAQS-revision, rather than conducting their own independent review.

The 2008/2011 Review of the Ozone Standard

As noted, EPA last changed the NAAQS for ozone in March 2008. Although the standard was strengthened, the level chosen at that time remains subject to controversy. The 23-member CASAC Review Panel unanimously recommended a range of standards more stringent than what the Administrator chose.[15]

With the change of Administrations in 2009, EPA agreed to reconsider the 2008 standard. As a result, a more stringent primary standard and a different version of the secondary standard were proposed in January 2010. After a year and a half of public comment and review, EPA sent what it considered a final set of standards to OMB for interagency review. The process was short-circuited, however, by a Presidential decision to await conclusion of the next regular review—the review now nearing completion—before promulgating any change.

The Primary Standard

The review completed in 2008 found evidence of health effects, including mortality, at levels of exposure below the then-current 0.08 ppm standard. As a result, both EPA staff and CASAC recommended strengthening the standard. CASAC stated: "There is no scientific justification for retaining the current [0.08 ppm] primary 8-hr NAAQS."[16] The panel unanimously recommended a range of 0.060 to 0.070 ppm (60 to 70 parts per billion) for the primary (health-based) 8-hour standard.

EPA staff also recommended strengthening the primary standard, in wording not quite so direct. The staff stated, "The overall body of evidence on ozone health effects clearly calls into question the adequacy of the current standard." They recommended "considering a standard level within the range of somewhat below 0.080 parts per million (ppm) to 0.060 ppm."[17]

Based on these recommendations, and his own judgment regarding the strength of the science, Stephen Johnson, the Bush Administration's last EPA Administrator, chose to finalize the standard at 0.075 ppm (75 parts per billion).[18] That revision led to designation of nonattainment areas in April and May 2012, as shown above in *Figure 1*.[19]

The Regulatory Impact Analysis that accompanied the 2008 standard identified 345 counties in 36 states in exceedance of the 0.075 ppm standard, using data for 2004-2006 (the most recent available at the time). By May 2012, when the nonattainment areas were actually designated, the number of counties in nonattainment had fallen to 221 in 27 states, based mostly on data for 2008-2010.[20] In the intervening years, emissions declined in most areas as more stringent standards for both mobile and stationary sources took effect. The recession may also have contributed to the lower numbers: when the economy is operating well below capacity, emissions generally decline.

Reconsideration of the 2008 Standard

As noted, EPA began a process to reconsider the 2008 ozone NAAQS in September 2009, and proposed a more stringent primary NAAQS in January 2010. The reconsideration process, which generally relied on the same data as that used to set the 2008 standard, led EPA to recommend a primary NAAQS of 0.070 ppm (70 ppb), within the range recommended by the CASAC Ozone Review Panel.

Although the draft final standard was withdrawn at the President's request before being promulgated, the agency subsequently made available documents prepared in support of its recommendation, including a Supplement to the 2008 Regulatory Impact Analysis.[21] The Supplemental RIA is discussed below, in the section on "Costs and Benefits of Control."

The Secondary Standard

As part of the review completed in 2008 and the 2010-2011 reconsideration process, EPA also assessed the secondary NAAQS for ozone. As explained above, secondary NAAQS are standards necessary to protect public welfare, a broad term that includes damage to crops, vegetation, property, building materials, climate, etc. Prior to 2008, the secondary standard was identical to the primary standard—0.08 ppm beginning in 1997.

Ozone affects both tree growth and crop yields, and the damage from exposure is cumulative over the growing season. In order to address this damage, EPA staff recommended in the 2008 review that the Administrator establish a new form for the secondary standard: a seasonal (three-month) average that would cumulate hourly ozone exposures for the daily 12-hour daylight window (termed a "W126 index"). The staff initially recommended a standard in a range of 7-21 parts per million-hours (ppm-hours). CASAC's ozone panel agreed unanimously that the form of the secondary standard should be changed as the staff suggested, but it did not agree that the upper bound of the range should be as high as 21 ppm-hours, suggesting that the upper bound be no higher than 15 ppm-hours.[22]

The Administrator's June 2007 proposal was in line with the staff recommendation, 7-21 ppm-hours, but his final March 2008 choice was to duplicate the primary standard he promulgated at that time. He set a secondary standard at 0.075 ppm averaged over 8 hours, rejecting the advice of both CASAC and his staff.

The secondary standard carries no deadline for attainment and has never been the subject of penalties or sanctions for areas that failed to meet it (unless they also violated a primary standard), but there was substantial disagreement between the Bush Administration EPA and the Office of Management and Budget (OMB) over the form in which this standard should be set.[23] Ultimately, OMB prevailed.[24]

Upon reconsideration, the 2011 draft standards would also have adopted the W126 index and would have set the secondary standard at 13 ppm-hours, in line with CASAC's recommendations. With the President's request to withdraw the draft standard and await completion of the current five-year review, a seasonal standard has yet to be implemented.

A total of 163 counties would have exceeded the draft seasonal standard sent to OMB in 2011.[25] All but 11 of these counties would also have exceeded the draft primary standard and would have had to address emissions of ozone precursors whether or not they exceeded the secondary standard.[26]

As discussed further below, however, few rural counties have ozone monitors, so the true extent of nonattainment with a seasonal secondary standard is currently unknowable. EPA proposed additional ozone monitoring requirements for both urban and non-urban areas in 2009;[27] as of May 2013, the agency had not issued a final monitoring rule.

CONTROLLING OZONE POLLUTION

Controlling ozone pollution is more complicated than controlling many other pollutants, because ozone is not generally emitted directly by pollution sources. Rather, it forms in the atmosphere when volatile organic compounds (VOCs) react with nitrogen oxides (NOx) in the presence of sunlight. The ozone concentration is as dependent on the temperature and amount of sunshine as it is on the presence of the precursor gases.

In general, ozone is a summertime pollutant. Other factors being equal, a cool, cloudy summer will produce fewer high ozone readings than a warm, sunny summer.

There are also complicated reactions that affect ozone formation. In general, lower emissions lead to less ozone, particularly lower emissions of VOCs. But under some conditions, *higher* emissions of NOx lead to lower ozone readings. This makes modeling ozone air quality and predicting attainment more difficult and contentious than the modeling of other air pollutants, and can affect consultations between EPA and the states to determine the boundaries of nonattainment areas and the adequacy of SIPs.

Most stationary and mobile sources are considered to be contributors to ozone pollution. Thus, there are literally hundreds of millions of sources of the pollutants of concern and control strategies require implementation of a wide array of measures. Among the sources of VOCs are motor vehicles (about 40% of total emissions), industrial processes, particularly the chemical and petroleum industries, and any use of paints, coatings, and solvents (about 40% for these sources combined). Service stations, pesticide application, dry cleaning, fuel combustion, and open burning are other significant sources of VOCs. Nitrogen oxides come overwhelmingly from motor vehicles and fuel combustion by electric utilities and other industrial sources.

WINTERTIME OZONE

An emerging set of issues has arisen in regard to wintertime ozone pollution in rural areas of the Western United States. Ozone is generally considered a summertime pollutant, but winter exceedances of the ozone NAAQS have recently been found to occur near oil and gas fields in rural areas of Wyoming, Utah, and Colorado. [28] At times, ozone concentrations as high as those in Los Angeles, the nation's smoggiest city, have been found in these areas—principally the Upper Green River Basin of Wyoming, the Uintah Basin of Utah, and a nearby area of Colorado.

The mechanism of ozone formation in the areas is still being studied, but it appears that the thousands of oil and gas wells in the two basins release volatile organic compounds (VOCs) that react with nitrogen oxides (NOx) from oil and gas operations and coal-fired power plants to create ozone. A study of the Uintah Basin by the National Oceanic and Atmospheric Administration, EPA, the Bureau of Land Management, the Western Energy Alliance, the Utah Department of Environmental Quality, and seven universities found that sources external to the basin are not major sources of the ozone found within it, and that among inventoried sources within the basin, 98% to 99% of the VOCs and 57% to 61% of the NOx comes from oil and gas operations.[29] The sunlight necessary for ozone to be created is magnified when it is reflected off of heavy snow cover. Snow cover also helps create temperature inversions that trap polluted air in the basins. In winters with little snow, there are few exceedances of the standards.[30]

EPA has recently promulgated standards requiring "green completions" to reduce emissions from oil and gas drilling. The impact of these regulations on wintertime ozone concentrations is yet to be determined.

COSTS AND BENEFITS OF CONTROL

As noted elsewhere in this report, EPA is prohibited by statute from taking cost into account in setting NAAQS; despite that prohibition, in order to comply with an executive order (E.O. 12866) and guidance from the Office of Management and Budget, the agency generally produces a Regulatory Impact Analysis (RIA) analyzing in detail the costs and benefits of new or revised NAAQS standards. The agency produced an RIA for its 2008 ozone NAAQS, and it released an 87-page draft supplement to that RIA in conjunction with the 2010-2011 reconsideration.[31] The supplemental RIA showed a range of estimates for three possible standards: 0.065 ppm; 0.070 ppm (the standard chosen by EPA at the conclusion of the reconsideration process, but not promulgated); and 0.075 ppm (the standard promulgated in 2008, and currently being implemented). EPA's estimates of the benefits and costs of the three options are summarized in *Table 1*.

The public health benefits of setting a more stringent ozone standard are the monetized value of such effects as fewer premature deaths, fewer hospital admissions, fewer emergency room visits, fewer asthma attacks, less time lost at work and school, and fewer restricted activity days.[32] The supplement to the RIA stated that the benefits of a 0.070 ppm primary standard would include the avoidance of 1,500 to 4,300 premature deaths annually in 2020.[33]

Other stated benefits of a 0.070 ppm standard in 2020 would include preventing the following, annually:

- 2,200 nonfatal heart attacks
- 6,600 hospital and emergency room visits
- 2,980 cases of acute or chronic bronchitis
- 44,000 cases of upper and lower respiratory symptoms
- 23,000 cases of aggravated asthma
- 770,000 days when people miss work or school
- 2.6 million days when people must restrict their activities.[34]

In the RIA supplement, the agency noted that "there are significant uncertainties in both cost and benefit estimates for the full range of standard alternatives."[35] Among the uncertainties are unquantified benefits (the effects of reduced ozone on forest health and agricultural productivity, for example) and unquantified disbenefits (reduced screening of UVB radiation and reduced nitrogen fertilization of forests and cropland).

The benefits will also vary, depending on which of the precursor pollutants nonattainment areas choose to control.

Table 1. Estimated Costs and Quantifiable Benefits of Ozone NAAQS Options, in 2020 (in billions of 2006 dollars)

Option	Costs	Quantifiable Benefits
0.075 ppm	$7.6 to $8.8	$6.4 to $15
0.070 ppm	$19 to $25	$11 to $31
0.065 ppm	$32 to $44	$19 to $53

Source: U.S. EPA, Draft Supplemental RIA, July 2011.

Notes: The data reflect annualized costs and annual monetized benefits of achieving the standard in 2020. Numerous benefits and disbenefits remain unquantified and are not included in the benefits estimates. The estimates assume a 7% discount rate for future costs and benefits and what EPA terms "a particular trajectory of aggressive technological change." According to the RIA, "An alternative storyline might hypothesize a much less optimistic technological trajectory, with increased costs, or with decreased benefits in 2020 due to a later attainment date."

The estimates include benefits of reduced fine particle (PM2.5) concentrations associated with ozone controls applied to meet each ozone standard option. Also, estimates only include areas assumed to meet the standard by 2020: they do not include the costs or benefits of attaining the alternate standards in the San Joaquin Valley and South Coast air basins in California, because the agency expects that nonattainment designations under the Clean Air Act for those areas would place them in categories afforded extra time beyond 2020 to attain the ozone NAAQS.

The RIA also states, "Of critical importance to understanding these estimates of future costs and benefits is that they are not intended to be forecasts of the actual costs and benefits of implementing revised standards."[36] Industry sources generally estimate the future costs of ozone NAAQS revisions to be greater than does EPA. EPA, on the other hand, maintains that its cost estimates are likely to overstate the eventual figures. In the agency's words, "Technological advances over time will tend to increase the economic feasibility of reducing emissions, and will tend to reduce the costs of reducing emissions."[37]

Benefits, both sides would agree, remain difficult to quantify, in part because of the difficulty of quantifying and valuing lives lost prematurely and other adverse health effects due to exposure to pollution.

ISSUES

The current ozone NAAQS review is likely to raise issues regarding the cost of compliance, the role of science in standard-setting, the interagency review process, the role of federal versus state and local pollution control measures, and the adequacy of the ozone monitoring network.

The Role of Cost

Because of its wide reach and potential cost, a proposed revision to the ozone NAAQS is likely to be among the most controversial rules EPA will propose in the coming year. In past reviews, industries that emit ozone precursors and broadly based business groups have complained forcefully that a more stringent standard would be too costly to attain.

The issue of cost is a perennial one in NAAQS decisions, even though EPA is prohibited by the Clean Air Act from considering costs in setting the standards. The Clean Air Act's §109 has been so interpreted since the NAAQS provisions were added to the act in 1970; in 2001, this interpretation was affirmed in a unanimous Supreme Court decision, *Whitman v. American Trucking Associations.*[38] The Court pointed to numerous other CAA sections where Congress had explicitly allowed consideration of economic factors, concluding that if Congress had intended to allow such factors in the setting of a primary NAAQS, it would have been more forthright— particularly given the centrality of the NAAQS concept to the CAA's regulatory scheme. The court concluded that §109(b)(1) "unambiguously bars cost considerations from the NAAQS setting process."[39]

This is not to say that cost considerations play no role in Clean Air Act decisions, including in *implementation* of a NAAQS. Cost-effectiveness is considered extensively by EPA and the states in selecting emission control options to meet the standards. Also, as discussed above, the agency prepares cost and benefit estimates at the time it proposes or promulgates a NAAQS, both for information purposes, and in order to comply with Executive Order 12866, under which the Office and Management and Budget (OMB) requires cost-benefit analysis of economically significant rules. But in deciding what level of ambient pollution poses a health threat, the statute bars consideration of costs.

Many in Congress would like to change this, by revising the CAA to require consideration of cost in NAAQS decisions. In the 112th Congress, the House twice passed legislation that would have done so: H.R. 2401 and H.R. 3409 would have required the EPA Administrator to take feasibility and costs into consideration in setting National Ambient Air Quality Standards. The Senate did not pass either bill. As of early May of this year, similar legislation had not been introduced in the 113th Congress, but proposal of more stringent standards could revive interest in the issue.

The Role of Science

Another major issue raised by proposed NAAQS standards may concern whether the Administrator's choices for the primary and secondary standards are backed by the scientific studies and recommendations of the agency's science advisers. One issue in recent years has concerned the form of the secondary ozone NAAQS: as discussed, the agency's staff and the independent CASAC review panel recommended a "W126" seasonal (rather than hourly) form of secondary ozone standard, to more adequately measure the impact of monitored ozone levels on forests and crops. This recommendation was dropped from the final decision in 2008. The W126 secondary standard was resurrected in the 2010-2011 reconsideration process, but it died a second time when the reconsidered standards as a whole were withdrawn.

More broadly, in the last two ozone NAAQS reviews, EPA (under both Republican and Democratic control) has been at odds with the Office of Management and Budget over the form and/or numerical level of the primary and secondary standards. As noted earlier, in 2008, OMB insisted on changes to EPA's recommended standards, at the completion of the multi-year agency review process. The result was promulgation of a secondary standard that was identical to the primary standard, rather than the new seasonal form of standard that the Administrator had chosen and that was unanimously recommended by the agency's science advisers.[40] In 2011, OMB and the President returned the draft final standard to EPA for "reconsideration," stating that the President did not support finalizing the rule "at this time."

Many in Congress and in the executive branch have decried the politicization of science in recent years. It has become a common complaint that EPA and other agencies that make regulatory decisions on scientific issues have done so based on false premises or on an inadequate review of the available research. The NAAQS process addresses this concern more directly than most aspects of the air pollution regulatory structure, by establishing an independent scientific advisory committee (CASAC) and by requiring the Administrator to discuss that committee's recommendations when a NAAQS standard is proposed or promulgated. But the Administrator remains free to disregard CASAC's conclusions, provided that the *Federal Register* notice summarizes any difference between the rule and CASAC's recommendation and explains the reasons for any difference.[41] This requirement has not prevented controversy over the scientific basis of NAAQS standards. It would seem likely that such controversy will arise again when the ozone NAAQS is proposed.

The Role of Federal Versus State and Local Pollution Control Measures

If the ozone NAAQS is made more stringent as a result of the current review, nonattainment areas will have to lower emissions sufficiently to comply with the new standard. This can be done through implementation of federal, state, and/or local measures. Current federal standards for cars, trucks, fuels, nonroad vehicles and engines, power plants, and other stationary pollution sources are reducing emissions, but at current levels, they are not strong enough to bring many areas into attainment, thus requiring state and local pollution control measures in addition. A more stringent NAAQS could also increase pressure on EPA to develop more stringent federal controls on the sources that contribute to ozone pollution.

A tighter ozone NAAQS might also provide an additional reason for Congress to revisit the issue of how to control emission sources that contribute to pollution downwind, in other states. Ozone, which forms in the atmosphere from chemical reactions downwind of the sources of precursor emissions, is the prime example of such downwind pollution. Under both the Bush and Obama Administrations, EPA has tried to address this interstate air pollution issue, but the D.C. Circuit Court of Appeals has overturned the agency's regulations.[42]

The agency is appealing the latest D.C. Circuit decision to the Supreme Court; but, absent a Supreme Court decision overturning the D.C. Circuit, the path forward is not clear. Congress could play a useful role in revisiting the issues posed by interstate air pollution.

Monitoring

The existing network of ozone monitors is concentrated in urban areas, because of the larger population potentially affected, and because most of the sources of ozone precursor emissions are located in such areas. But, as noted earlier, ozone is not generally emitted directly by pollution sources. It forms in the atmosphere downwind of emission sources. Thus, rural areas can have high ozone concentrations, unless they are located a substantial distance from any urban area or other source of emissions. In addition to the potential health impacts of ozone, the new form of the secondary NAAQS discussed by EPA suggests a need for additional monitoring in rural areas to detect impacts on forests and agricultural production. Both EPA and state monitoring budgets are constrained, however, raising questions as to how any additional monitoring requirement would be funded.

The agency has, in a separate rulemaking, proposed changing the minimum monitoring requirements for both urban and non-urban areas.[43] That proposal would require that each state operate at least three ozone monitors in non-urban areas. It would also require at least one ozone monitor in each urban area with a population between 50,000 and 350,000. Nearly four years after proposal, the requirements have not been finalized.

EPA has recently decided to add to its compliance database data from a small monitoring network previously used only for research, the Clean Air Status and Trends Network (CASTNET). This decision has proved controversial in at least one state, which reportedly decided to shut down a CASTNET monitor that was showing nonattainment. The CASTNET monitors, many of which are in rural areas, may be indicators of a broader ozone problem that lack of monitoring has failed to identify.[44] Without additional monitors, this problem will remain largely unaddressed.

End Notes

[1] Availability of the assessment was announced at 78 Federal Register 11172, February 15, 2013. For a summary, see "EPA Assessment Underscores Relationship Between Ozone Exposure, Respiratory Effects," Daily Environment Report, February 19, 2013. The assessment itself is 1,251 pages long and is available at http://www.epa.gov/ncea/isa/ ozone.htm.

[2] U.S. EPA, "Supplement to the Regulatory Impact Analysis for Ozone," Fact Sheet, January 7, 2010, at http://www.epa.gov/glo/pdfs/fs20100106ria.pdf.

[3] U.S. EPA, Regulatory Impact Analysis, Final National Ambient Air Quality Standard for Ozone, July 2011, p. 64, at http://www.epa.gov/glo/pdfs/201107_OMBdraft-OzoneRIA.pdf.

[4] Ibid., p. 74.

[5] U.S. EPA, "Final Revisions to the National Ambient Air Quality Standards for Ozone," Fact Sheet, March 12, 2008, at http://www.epa.gov/glo/pdfs/2008_03_factsheet.pdf.

[6] For specifics, see EPA's "Green Book," at http://www.epa.gov/oaqps001/greenbk/hntc.html.

[7] A listing of these areas, as well as areas still designated nonattainment for the 1997 standard, can be found at http://www.epa.gov/oaqps001/greenbk/o8index.html.

[8] The Clean Air Act's definition of welfare is found in Section 302(h) of the act (42 U.S.C. 7602).

[9] Authority to establish NAAQS comes from both Sections 108 and 109 of the act; this definition of criteria pollutants is found in Section 108. The authority and procedures for controlling the sources of criteria pollutants are found throughout Titles I, II, and IV of the act. Pollutants that are less widely emitted are generally classified as "hazardous air pollutants" and are regulated under a different section of the act (Section 112).

[10] For a more detailed discussion, see CRS Report RL30853, Clean Air Act: A Summary of the Act and Its Major Requirements.

[11] The schedule was set by a consent decree that settled a lawsuit filed by the American Lung Association (American Lung Association v. Leavitt, D.D.C., No. 03-778, modified consent decree approved 12/16/04). EPA agreed that it would propose whether to retain or revise the ozone standard by June 20, 2007, and take final action by March 12, 2008.

[12] Mississippi v. EPA, D.C. Cir., No. 08-1200, notice filed 9/16/09.

[13] "Statement by the President on the Ozone National Ambient Air Quality Standards," September 2, 2011, at http://www.epa.gov/glo/actions.html.

[14] U.S. EPA, Office of Research and Development, Integrated Science Assessment of Ozone and Related Photochemical Oxidants, Final Report, February 2013, at http://www.epa.gov/ncea/isa/ozone.htm.

[15] In a letter to the Administrator sent after promulgation of the NAAQS, the panel's chair stated, "Nevertheless, the members of the CASAC Ozone Review Panel do not endorse the new primary ozone standard as being sufficiently protective of public health. The CASAC—as the Agency's statutorily-established science advisory committee for advising you on the national ambient air quality standards—unanimously recommended decreasing the primary standard to within the range of 0.060–0.070 ppm. It is the Committee's consensus scientific opinion that your decision to set the primary ozone standard above this range fails to satisfy the explicit stipulations of the Clean Air Act that you ensure an adequate margin of safety for all individuals, including sensitive populations." Letter of Rogene F. Henderson, Chair, Clean Air Scientific Advisory Committee, to EPA Administrator Stephen L. Johnson, April 7, 2008, at http://yosemite.epa.gov/sab/sabproduct. nsf/4AF8764324331288852574250069E494/$File/EPA-CASAC-08-009- unsigned.pdf.

[16] Letter of Rogene Henderson, Chair, Clean Air Scientific Advisory Committee, to Hon. Stephen L. Johnson, EPA Administrator, October 24, 2006, available at http://yosemite.epa.gov/sab/sabproduct.nsf/ AB290E0DB8B72A33852572120055858F/$File/casac-07-001.pdf.

[17] "Review of National Ambient Air Quality Standards for Ozone Final Staff Paper, Human Exposure and Risk Assessments and Environmental Report," Fact Sheet, at http://www.epa.gov/ttn/naaqs/standards/ozone/data/ 2007_01_finalsp_factsheet.pdf.

[18] All of EPA's references to the standard are expressed as parts per million (e.g., 0.075 ppm), but many references in the press convert this to a more readable parts per billion (i.e., 75 parts per billion). In order to avoid confusion when quoting from EPA sources, this report generally uses the more cumbersome parts-per-million form.

[19] Detailed information on the designations, including links to Federal Register notices, can be found on EPA's website at http://www.epa.gov/airquality/ozonepollution/designations/ 2008standards/regs.htm#may12.

[20] A few states had certified monitoring data for 2009-2011, and their designations were based on that three-year period.

[21] See "EPA Releases Documents on the National Ambient Air Quality Standard for Ozone," at http://www.epa.gov/ glo/actions.html.

[22] Letter of Rogene Henderson, Chair, Clean Air Scientific Advisory Committee, to Hon. Stephen L. Johnson, EPA Administrator, March 26, 2007, p. 3, available at http://yosemite. epa.gov/sab/sabproduct.nsf/ FE915E916333D776852572AC007397B5/$File/casac-07-002.pdf.

[23] OMB maintained that EPA had failed to consider or evaluate the effects of a W126 standard on "economic values, personal comfort, and well-being"—terms that are also included in the Clean Air Act's definition of welfare—and thus did not provide a balanced consideration of welfare effects, as required by the act. OMB also maintained that EPA had not adequately demonstrated that the proposed secondary standard would be more protective than one set equal to the primary standard. See "Ozone Secondary Standard," Memorandum of Marcus Peacock, EPA Deputy Administrator, to Susan Dudley, OMB, March 7, 2008, p. 2, at http://oversight-archive.waxman.house.gov/documents/ 20080520091448.pdf. The House Oversight and Government Reform Committee made this and other relevant documents from the 2008 review available, two months after the review's completion. See House Oversight and Government Reform Committee, "Supplemental Information on the Ozone NAAQS," Memorandum from Majority Staff to Members of the Committee, May 20, 2008, in White House Overruled EPA Administrator on Ozone Regulation, at http://oversight-archive.waxman.house.gov/story.asp?id=1958. For additional discussion, also see CRS Report RL34057, Ozone Air Quality Standards: EPA's March 2008 Revision, by James E. McCarthy.

[24] Letter of Susan E. Dudley, Administrator, Office of Information and Regulatory Affairs, OMB, to Stephen L. Johnson, EPA Administrator, March 13, 2008, at http://oversight-archive.waxman.house.gov/documents/ 20080520092019.pdf.

[25] For additional information, including a map of the potential nonattainment areas for the draft secondary standard, see U.S. EPA, Regulatory Impact Analysis, Final National Ambient Air Quality Standard for Ozone, July 2011, pp. 48-49, at http://www.epa.gov/glo/pdfs/201107_ OMBdraft-OzoneRIA.pdf.

[26] The 11 counties that would have exceeded only the draft secondary standard were in Arizona, Colorado, New Mexico, and Utah. Supplemental RIA, p. 49.

[27] For information on the proposed monitoring rule, see "Fact Sheet: Proposal to Revise the National Ambient Air Quality Standards for Ozone," p. 4, at http://www.epa.gov/air/ozone pollution/pdfs/fs20100106std.pdf.

[28] Wintertime ozone levels are occasionally elevated in urban areas, as well, when concentrations of pollution become trapped in cold stagnant air near the Earth's surface. What is unusual about the rural areas discussed in this section of the report is that they do not experience high ozone concentrations in warm weather. In addition, elevated ozone in Western rural areas is a newly observed phenomenon, whereas pollution episodes associated with temperature inversions in urban areas have been observed for decades.

[29] 2012 Uintah Basin Winter Ozone & Air Quality Study, Final Report, February 1, 2013, p.2, at http://rd.usu.edu/files/ uploads/ubos_2011-12_final_report.pdf. The remaining NOx comes primarily from the Bonanza Power Plant.

[30] In the Uintah Basin, for example, winter 2011-2012 measurements indicated no exceedance of the 75 ppb NAAQS for ozone: the highest 8-hour average reading was 63 ppb. In 2010-2011, by contrast, there were 25 winter days with ozone levels exceeding the 75 ppb standard, with the highest 8-hour reading being 139 ppb.

[31] Links to the RIAs and other information related to the 2008 and draft 2011 ozone NAAQS can be found at http://www.epa.gov/glo/actions.html.

[32] For a full discussion of these variables and their monetized values, see Chapter 6 of the 2008 RIA at http://www.epa.gov/ttn/ecas/regdata/RIAs/452_R_08_003.pdf.

[33] U.S. EPA, Draft Supplemental RIA, p. 7.

[34] Ibid.

[35] Ibid., p. 17.

[36] Ibid.

[37] Ibid., p. 19.

[38] 121 S. Ct. 903 (2001).

[39] For further discussion of the American Trucking case, see CRS Report RS20860, The Supreme Court Upholds EPA Standard- Setting Under the Clean Air Act: Whitman v. American Trucking Ass'ns, by Robert Meltz and James E. McCarthy.

[40] See the section on The Secondary Standard and footnote 23, above.

[41] Section 307(d)(3) and (d)(6).

[42] For a discussion of the Bush and Obama Administration regulations and the D.C. Circuit decisions remanding them, see CRS Report R42895, Clean Air Issues in the 113th Congress: An Overview.

[43] 74 Federal Register 34525, July 16, 2009.

[44] For additional information, see "States Scramble as EPA Shifts Research Monitors into Regulatory Mode," Greenwire, April 30, 2013.

In: Air Quality Observation in the US
Editor: Malcolm Parisi
ISBN: 978-1-63117-154-3

Chapter 3

AIR QUALITY: EPA'S 2013 CHANGES TO THE PARTICULATE MATTER (PM) STANDARD*

Robert Esworthy

SUMMARY

On January 15, 2013, the Environmental Protection Agency (EPA) published a final rule revising the National Ambient Air Quality Standard (NAAQS) for particulate matter (PM). The revised air quality standards were completed pursuant to the Clean Air Act (CAA) and, in part, in response to a court order and consent agreement. Based on its review of scientific studies available since the agency's previous review in 2006, EPA determined that evidence continued to show associations between particulates in ambient air and numerous significant health problems, including aggravated asthma, chronic bronchitis, non-fatal heart attacks, and premature death. Populations shown to be most at risk include children, older adults, and those with heart and lung disease, and those of lower socioeconomic status. EPA's review of and revisions to the PM NAAQS have generated considerable debate and oversight in Congress.

---

* This is an edited, reformatted and augmented version of a Congressional Research Service Publication, CRS Report for Congress R42934, dated March 12, 2013.

The January 2013 revisions change the existing (2006) annual health-based ("primary") standard for "fine" particulate matter 2.5 micrometers or less in diameter (or $PM_{2.5}$), lowering the allowable average concentration of $PM_{2.5}$ in the air from the current level of 15 micrograms per cubic meter ($\mu g/m^3$) to a limit of 12 $\mu g/m^3$. The annual $PM_{2.5}$ NAAQS is set so as to address human health effects from chronic exposures to the pollutants. The existing "24-hour primary standard" for $PM_{2.5}$ that was reduced from 65 $\mu g/m^3$ to 35 $\mu g/m^3$ in 2006 was retained, as was the existing standard for larger, but still inhalable, "coarse" particles less than 10 micrometers in diameter, or PM_{10}. "Secondary" standards that provide protection against "welfare" (non-health) effects, such as ecological effects and material deterioration, are identical to the primary standards and the same as in 2006. The proposed rule published June 29, 2012, solicited comments on two options for a 24-hour $PM_{2.5}$ standard to improve visibility that were not adopted in the final rule.

EPA revised the Regulatory Impact Analysis (RIA) accompanying its June 2012 proposed rule in part in response to comments received regarding the agency's cost and benefit estimates. In its December 2012 RIA, EPA estimated that the potential "quantifiable" health benefits (2010 \$) associated with attaining the PM standard would range from \$4.0 billion to \$9.1 billion, and costs (2010 \$) would range from \$53.0 million to \$353.0 million. Some stakeholders and some Members continue to express concerns that cost impacts would be more significant than those estimated by EPA for those areas out of compliance with the new standards.

EPA's revisions to the PM NAAQS do not directly regulate emissions from specific sources, or compel installation of any pollution control equipment or measures, but indirectly could affect operations at industrial facilities and other sources throughout the United States. Revising PM NAAQS starts a process that includes a determination of areas in each state that exceed the standard and must, therefore, reduce pollutant concentrations to achieve it. Following determinations of these "nonattainment" areas based on multiple years of monitoring data and other factors, state and local governments must develop (or revise) State Implementation Plans (SIPs) outlining measures to attain the standard. These often involve promulgation of new regulations by states, and the issuance of revised air permits. The process typically takes several years. Based on statutory scheduling requirements, nonattainment designations for revised PM NAAQS would not be determined until the end of 2014, and states would have until at least 2020 to achieve compliance with the January 2013 revised $PM_{2.5}$ NAAQS.

INTRODUCTION

On January 15, 2013, the Environmental Protection Agency (EPA) published a final rule in the *Federal Register* to strengthen the National Ambient Air Quality Standard (NAAQS) for particulate matter (PM),[1] intended to address potential health effects (including chronic respiratory disease and premature mortality) associated with short- and long-term exposure to particulate matter pursuant to the Clean Air Act (CAA).[2] The CAA, enacted in 1970 and amended in 1990, requires EPA to set minimum NAAQS for six "criteria air pollutants, including, ozone ('smog'), particulate material ('soot'), sulfur dioxides, nitrogen oxides, carbon monoxide, and lead." The law also requires EPA to evaluate each NAAQS every five years to determine whether it is adequately protective of human health and the environment, based on the most recent science.[3]

The EPA Administrator signed the final PM NAAQS rule on December 14, 2012, as per a June 6, 2012, order issued by the U.S. Court of Appeals for the District of Columbia Circuit in response to petitions filed by advocacy groups and 11 states,[4] and as agreed to in a September 4, 2012, consent decree.[5] EPA's most recent statutorily required review and proposal has generated controversy and national debate among a variety of stakeholders including industry groups, health and environmental advocacy groups, and states, as well as oversight in Congress. Similar controversy and debate transpired during the previous changes leading up to the existing PM NAAQS promulgated October 2006, and those established in 1997.

EPA published a proposed rule on June 29, 2012,[6] which started a nine-week public comment period that ran through August 31, 2012. EPA also held two public hearings for the proposal on July 17, 2012, in Philadelphia, PA, and July 19, 2012, in Sacramento, CA.[7] EPA reportedly received and considered more than 230,000 written comments in determining the final PM standard.

The June 2012 proposal and the January 2013 final PM NAAQS rule were the culmination of EPA's statutorily required review of the NAAQS under the CAA based on studies available through mid-2009 and recommendations of EPA staff and a scientific advisory panel (Clean Air Scientific Advisory Committee, or CASAC)[8] established by the CAA.[9] The agency initiated the review not long after the 2006 promulgation of the PM NAAQS.[10] EPA staff reassessed scientific studies considered in setting the 2006 PM NAAQS revisions, reviewed and analyzed extensive subsequent research, and considered public comments and recommendations of the CASAC.

Based on the scientific evidence and comments considered, EPA Administrator Lisa P. Jackson signed the final rule that would change the current standard primarily by lowering the annual health-based ("primary") standard for fine particles smaller than 2.5 microns ($PM_{2.5}$). In the final rule, the "secondary" standards that provide protection against "welfare" (non-health) effects, such as ecological effects and material deterioration, are identical to the primary standards, the same as in 2006. The final rule relies on the existing secondary 24-hour standard to protect against visibility impairment, and did not adopt a separate standard included among options in the June 2012 proposal. Also, as proposed,[11] the final rule did not modify the standards for inhalable "coarse" particles larger than 2.5 but smaller than 10 microns (PM_{10}). Some stakeholders in the agricultural community and some Members maintained a particular interest in EPA's consideration of the PM_{10} standards and potential impacts that revising the NAAQS may impose on the agricultural operations.[12]

As per statutory scheduling requirements under the CAA, the final designation of areas (primarily counties) as nonattainment for any revised PM standards would not be determined until the end of 2014, and states would have until at least 2020 to achieve compliance with the January 2013 PM NAAQS. In its revised Regulatory Impact Analysis (RIA) accompanying the final rule assessing the costs and benefits of proposed revisions to the PM NAAQS, EPA estimated that tightening the $PM_{2.5}$ annual standard would add further health benefits beyond those anticipated with the promulgation of the 2006 PM NAAQS.[13] Others have suggested that potential health benefits of tightening the PM NAAQS might be higher than EPA's estimates.[14] On the other hand, tighter standards could impose additional compliance requirements on communities, states, industry, and others, at what some stakeholders and Members contend will be a substantial economic cost. EPA expects that requirements and emission reductions associated with existing and recently promulgated federal regulations under the CAA will significantly allay impacts of complying with the revised PM standards, and anticipates that virtually all counties will meet the standards as promulgated in 2020.

Several recent and pending EPA regulations implementing the various pollution control statutes enacted by Congress garnered vigorous oversight during the 112th Congress.[15] Members expressed concerns in hearings, through bipartisan letters commenting on proposed regulations, and through introduced legislation that would delay, limit, or prevent certain EPA actions. Particular attention was focused on EPA's implementation of the CAA. Because of health and cost implications, NAAQS decisions historically have been the source of significant concern to some in Congress. The evolution and development of the PM NAAQS, in particular, have been the subject of extensive oversight. During the 112th Congress, some Members expressed concerns in hearings, letters to the Administrator, and proposed legislation in anticipation of potential changes to the PM NAAQS, and the January 2013 final rule is expected to generate further oversight. Some Members[16] and industry stakeholders had urged EPA to delay the final rule, while conversely, others, including some states[17] and various environmental and public health advocacy groups, urged timely completion of a tighter standard. Changes to the NAAQS historically have triggered litigation alleging the standards are too stringent or not stringent enough, and often resulted in delays in implementation.

This CRS report summarizes EPA's January 15, 2013, final and June 2012 proposed changes to the PM NAAQS and includes comparisons with previous (1997 and 2006) promulgated and proposed standards. Key actions leading up to the agency's determination, and potential issues and concerns associated with changing the $PM_{2.5}$ annual standard, are also highlighted.

BACKGROUND

Particulate matter is one of six "criteria pollutants" for which EPA has promulgated NAAQS under the CAA.[18] The others are ozone ("smog"), nitrogen oxides (NO_x),[19] sulfur oxides (SO_x, or, specifically, SO_2), carbon monoxide (CO), and lead (Pb).

$PM_{2.5}$ can be emitted directly from vehicles, smokestacks, and fires but can also form in reactions in the atmosphere from gaseous precursors, including sulfur oxides, nitrogen oxides, and volatile organics occurring naturally or as emissions typically associated with gasoline and diesel engine exhaust, and from utility and other industrial processes. PM_{10} (or coarse PM) is an indicator used in the NAAQS to provide protection from slightly larger (in the range of 2.5 to 10 microns or thoracic "coarse" particles), but still inhalable particles that penetrate into the trachea, bronchi, and deep lungs. These particles are often associated with dust from paved and unpaved roads, construction and demolition operations (including mining), and sometimes with certain industrial processes and agriculture operations, as well as biomass burning.

Establishing NAAQS does not directly limit emissions; rather, it represents the EPA Administrator's formal judgment regarding the concentration of a pollutant in ambient air that will protect public health with an "*adequate margin of safety*." Under Sections 108-109 of the CAA,[20] Congress mandated that EPA set national ambient (outdoor) air quality standards for pollutants whose emissions "may reasonably be anticipated to endanger public health (primary standards) or welfare[21] (secondary standards)" and "the presence of which in the ambient air results from numerous or diverse mobile or stationary sources." The process for setting and revising NAAQS consists of the statutory steps incorporated in the CAA over a series of amendments. Several other steps have also been added by EPA, by executive orders, and by subsequent regulatory reform enactments by the Congress.

Section 109(d)(1)) of the CAA requires EPA to review the criteria that serve as the basis for the NAAQS for each covered pollutant every five years, to either reaffirm or modify previously established NAAQS. Prior to the January 2013 revisions, EPA has revised the PM NAAQS three times, in 1987, 1997, and October 2006, to ensure that the standards continue to provide adequate protection for public health and welfare.[22]

A February 24, 2009, decision by the U.S. Court of Appeals for the District of Columbia Circuit had remanded elements of EPA's decisions as promulgated in October 2006, in particular the decision not to tighten the primary annual NAAQS for $PM_{2.5}$, to the agency for further consideration but did not vacate the revised standard nor set a specific timeline. The decision was in response to petitions filed in the D.C. Circuit by 13 states, industry, agriculture, business, and environmental and public health advocacy groups, challenging certain aspects of EPA's revisions for both $PM_{2.5}$ and PM_{10}. The D.C. Circuit granted the petitions in part with regard to the $PM_{2.5}$ annual standard and the secondary standards for $PM_{2.5}$ and PM_{10} (including visibility impairment), denying other challenges.[23]

Concerned with delays in EPA's schedule for proposing revisions to the 2006 PM NAAQS, the American Lung Association and the National Parks Conservation Association, and nine states separately filed petitions with the D.C. Circuit in November 2011 urging the court to order EPA's immediate compliance with the February 2009 remand. Subsequently, in February 2012 the two organizations sued EPA in the D.C. Circuit for failing to fulfill their statutory duty to review the October 2006 PM NAAQS within five years,[24] and a coalition of 11 states filed a similar suit with the U.S. District Court Southern District of New York.[25] In response, the D.C. Circuit initially directed EPA to sign a proposed rule concerning its decision regarding revisions to the PM

NAAQS by June 7, 2012, and following a motion filed by the agency, amended the deadline to June 14, 2012.[26] As part of a September 4, 2012, consent decree, EPA agreed to finalize revisions to the PM NAAQS by December 14, 2012.[27]

Promulgation of a revised NAAQS, such as the PM NAAQS, initiates a series of statutorily required actions, ultimately culminating in issuance of permits pursuant to state regulations in a State Implementation Plan (SIP). The first step is designation of attainment and nonattainment areas, based on the accumulated results of ambient air monitoring and modeling data. States first propose to designate certain geographic areas (e.g., counties) as either "attainment" or "nonattainment," depending on whether the data indicate the concentrations of pollutants will be below or above the NAAQS. After extensive dialogue with state officials, EPA either approves the proposed attainment and nonattainment areas, or sends back to states proposed revisions. EPA and states generally come to an agreement about the area designations. Following this designation, approved by EPA, states then develop a SIP, which consists essentially of state regulations to be implemented by states that would affect the state emissions inventory, and therefore the expected or modeled concentrations of air pollutants. After approval of the SIP as being adequate to control air pollution and reduce the ambient air pollutant concentrations in designated nonattainment areas, the states then issue permits (new or modified) for facilities whose emissions affect the air in designated nonattainment areas.

EPA'S JANUARY 2013 FINAL CHANGES TO THE PM NAAQS

EPA's 1997 revisions to the PM NAAQS[28] revised the standards focused on particles smaller than 10 microns (PM_{10} or coarse particles) established in 1987,[29] and introduced standards for "fine" particles smaller than 2.5 microns ($PM_{2.5}$) for the first time. The primary (health protection) PM NAAQS as revised in 2006 include an *annual* and a *daily* (24-hour) limit for $PM_{2.5}$, but only a daily limit for PM_{10}. To attain the $PM_{2.5}$ annual standard, the three-year average of the weighted annual arithmetic mean $PM_{2.5}$ concentration at each monitor within an area must not exceed the maximum limit set by the agency. The 24-hour standards are a concentration-based percentile form,[30] indicating the percent of the time that a monitoring station can exceed the standard. For instance, a 98th percentile 24-hour standard indicates that a monitoring station can exceed the standard 2% of the time during the year. For $PM_{2.5}$ and PM_{10}, the secondary NAAQS, which are set at a level "requisite to protect the public welfare," are the same as the primary standards.

In the final rule published by EPA on January 15, 2013, the $PM_{2.5}$ and PM_{10} standards and other implementation changes are as follows:[31]

Primary (Public Health) PM Standards

- **$PM_{2.5}$:** EPA revised the *annual* standard, which currently is 15 micrograms per cubic meter ($\mu g/m^3$), by setting a new limit of 12 $\mu g/m^3$ (the proposal included an optional limit of 13 $\mu g/m^3$ and solicited comment for 11 $\mu g/m^3$); compliance with the "annual" standard is determined by whether the three-year average of its annual average $PM_{2.5}$ concentration (at each monitoring site in the area) is less than or equal to 12 $\mu g/m^3$; as proposed, EPA retained the *daily* (24-hour) standard at 35 $\mu g/m^3$ based on the current three-year average of the 98th percentile of 24- hour $PM_{2.5}$ concentrations as established in 2006.
- **PM_{10}:** As proposed, EPA retained the current *daily* standard of no more than one exceedance of concentrations of 150 $\mu g/m^3$ per year on average over three years; there is no current *annual* standard for PM_{10} (the previous annual maximum concentration standard of 50 $\mu g/m^3$ was eliminated by EPA in 2006).[32]

Secondary (Welfare) PM Standards

- **$PM_{2.5}$ and PM_{10}:** As proposed, secondary (welfare) NAAQS are the same as the primary standards, the same correlations as the 2006 PM NAAQS, with the exception of visibility impairment associated with $PM_{2.5}$.
- **$PM_{2.5}$ Visibility Impairment:** The final rule did not add a distinct secondary standard as proposed, defined in terms of a $PM_{2.5}$ visibility index based on speciated[33] $PM_{2.5}$ mass concentrations and relative humidity data to calculate light extinction on a deciview (dv) scale[34] similar to the current Regional Haze Program.[35] Specifically, the proposal would have set a 24-hour averaging time of 30 or 28 deciviews (dv) based on a 90th percentile form over three years. EPA also sought comment on alternative levels (down to 25 dv) and

averaging times (e.g., 4 hours). Based on public comment and further analysis of air quality monitoring data, EPA concluded that the current secondary standard would provide visibility protection greater than or equal to 30 dv.[36]

Implementation Changes

- **Monitoring:**[37] As proposed, updates several aspects of monitoring regulations including requiring relocating a small number of $PM_{2.5}$ monitors[38] to be collocated with measurements of other criteria pollutants (e.g., nitrogen dioxide (NO_2) and carbon monoxide (CO)) near-roadway monitoring so as to ensure these monitors are at one location in each urban area with a population of 1 million or more, and to be phased in starting with the largest areas (2.5 million or more populations) by January 1, 2015, and extended to the remainder of areas by January 1, 2017. Includes the use data from existing Chemical Speciation Network or the EPA/National Park Service IMPROVE monitoring network to determine whether an area meets the proposed secondary visibility index standard for $PM_{2.5}$. No changes to PM_{10} monitoring.
- **Air Quality Index (AQI):** As proposed, updates the AQI (EPA's color-coded tool for informing the public about air quality and associated measures for reducing risks of exposure) for $PM_{2.5}$ by changing the upper end range for "Good" category (an index value of 50) on the overall scale (0 to 500 based on conversion of $PM_{2.5}$ concentrations) to the level of the revised annual $PM_{2.5}$ standard (12 $\mu g/m^3$). Also as proposed, EPA is setting the 100 value of the index scale ("Moderate") at the level of the current 24-hour $PM_{2.5}$ standard, which is 35 $\mu g/m,^3$ and the AQI of 150 ("Unhealthy Sensitive Groups") at 55 $\mu g/m^3$. The current upper ends for the "Hazardous" (500), "Unhealthy" (200) and "Very Unhealthy" (300) AQIs are retained.[39]
- **Prevention of Significant Deterioration (PSD):**[40] EPA revised the PSD permitting program (rules) with respect to the revised PM NAAQS so as not to "unreasonably delay" pending permits and establish a "grandfather" provision for permit applications if: the permitting agency deems an application complete by December 14, 2012; or public notice for a draft permit or preliminary determination has been published (for public comment) no later than the effective

date of revised PM NAAQS (60 days after January 15, 2013, publication in the *Federal Register*). This provision would not apply to NAAQS for other criteria pollutants and permits not meeting these criteria would have to demonstrate compliance with the revised standards once they are finalized.

Comparison of the January 2013 Revised $PM_{2.5}$ Standards with Previous Promulgated and Proposed Alternative PM Standards

The final $PM_{2.5}$ daily standard established in 2006 was among the less stringent within the range of alternative levels recommended by EPA staff, and the annual standard is not as stringent as the standard recommended by the CASAC. The decision to retain the annual $PM_{2.5}$ standard was also less than recommended. **Table 1** below shows the January 2013 revised $PM_{2.5}$ annual standard in comparison to the June 2012 proposed options and to the annual and daily standards for 1997 and 2006 promulgated standards, and alternative levels recommended prior to the 2006 final revisions.

REVIEW PROCESS LEADING UP TO THE JANUARY 2013 REVISED PM NAAQS

The CAA as enacted includes specific requirements for a multistage process to ensure the scientific integrity under which NAAQS are set, laying the groundwork for the Administrator's determination of the standard, and the procedural process for promulgating the standard.[41] Primary NAAQS, as described in Section 109(b)(1), were to be "ambient air quality standards the attainment and maintenance of which in the judgment of the Administrator, based on such criteria and allowing an adequate margin of safety, are requisite to protect the public health."

Based on this premise, the CAA specifies the criterion to be used by the Administrator in deciding on the final standard, including preparation of a "criteria document" that summarizes scientific information assessed. The act also requires the establishment and role of an independent advisory committee (CASAC)[42] to review EPA's supporting scientific documents, and the timeline for completing specific actions. EPA administratively added the preparation of a "staff paper" that summarizes the criteria document and lays out policy options. This EPA document typically serves as the basis for CASAC review and comment. EPA revised certain aspects (not including reinstating the closure letter) of the CASAC review process most recently in May 2009.[43] In addition, Executive Order 12866 requires a Regulatory Impact Analysis (RIA), although the economic impact analysis is essentially only for informational purposes and cannot be directly considered as part of the decision in determining the NAAQS.[44]

Table 1. Promulgated, Proposed, and Alternative $PM_{2.5}$ Primary (Health) National Ambient Air Quality Standards (NAAQS)

$PM_{2.5}$ NAAQS Options	**24-Hour Primary**	**Annual Primary**
	micrograms per cubic meter = μg/m^3	
1997 Promulgated PM NAAQS	65 μg/m^3	15 μg/m^3
CASAC Recommendation (June 2005)	35-30 μg/m^3	14-13 μg/m^3
EPA Final "Staff Paper" (Dec. 2005)	35-25 μg/m^3	15 μg/m^3
	or	
	40-30 μg/m^3	14-12 μg/m^3
Dec. 2005 Proposed PM NAAQS Rule	35 μg/m^3	15 μg/m^3
2006 Promulgated PM NAAQS	35 μg/m^3	15 μg/m^3
CASAC Recommendation (August 2010)	35-30 μg/m^3	13-11 μg/m^3
EPA Final "Staff Paper" (April 2011)	35-30 μg/m^3	13-11 μg/m^3
2012 Proposed Rule (June 2012)	35 μg/m^3	13-12 μg/m^3 (EPA also solicited comments for a limit of 11 μg/m^3)
January 15, 2013, Final Rule	35 μg/m^3	12 μg/m^3

Source: Prepared by the Congressional Research Service (CRS) with information from EPA's January 15, 2013, final rule and June 2012 proposal and related technical documents, and the December 2006 promulgated PM NAAQS and supporting technical and policy documents (http://www.epa.gov/air/particles/actions.html).

Note: $PM_{2.5}$ = "fine" particulate matter 2.5 micrometers or less in diameter.

Beginning June 2007 with its general call for information,[45] EPA initiated the current PM NAAQS review, which culminated in assessments of the scientific research and risk analyses, and ultimately the April 2011 publication of the staff's final *Policy Assessment for the Review of the Particulate Matter National Ambient Air Quality Standards (or PM Policy Assessment).*[46] The staff paper presented the staff conclusions and recommendations on the elements of the PM standard based on evaluation of the policy implications of the scientific evidence contained in the criteria document and the results of quantitative analyses (e.g., air quality analyses, human health risk assessments, and visibility analyses) of that evidence. **Table B-1** in **Appendix B** provides a chronological listing of EPA's supporting documents leading up to the June 2012 proposed PM NAAQS.

Supplemental to public comments solicited in the *Federal Register*, the CASAC reviewed EPA's drafts and final documents supporting the science and policy behind the Administrator's decisions in the June 2012 PM NAAQS proposal. The CASAC conducted meetings and consultations, and submitted written overviews, providing their views of the validity and completeness of the agency's assessments and findings, and recommending improvements. CASAC's final product, its review of EPA's second external review draft of the "PM Policy Assessment," was completed June 2010.[47]

Table B-2 in **Appendix B** provides a chronological summary of CASAC consultations and reviews of the supporting documents for the June 2012 proposal.

The April 2011 EPA policy assessment ("staff paper") concluded, and the CASAC panel concurred in its final recommendations, that the scientific evidence supported modifying the $PM_{2.5}$ primary standard and considering options for revising the secondary standard for reducing visibility impairment associated with PM. Recognizing certain limitations of the data, the policy assessment included a range of alternatives for consideration by the Administrator for modifying the current PM NAAQS. These recommendations were the core basis for the June 2012 proposal[48] and the Administrator's final decision to revise the PM NAAQS, taking into account other factors including public comments received in response to the June 2012 proposal.

The EPA staff paper included possible modifications to strengthen certain aspects of the PM_{10} standard. However, staff and CASAC placed considerable emphasis on continuing uncertainties and lack of sufficient data to initiate relevant quantitative risk assessment to support such modifications to the standard. As presented in the June 2012 *Federal Register* notice, the Administrator provisionally concluded that the growing evidence continued to support the appropriateness of the existing primary 24-hour PM_{10} standard's protection of short-term health effects, and proposed to retain the existing PM_{10} standard.[49]

A perennial issue in conducting NAAQS reviews is whether the agency is basing its decisions on those studies that reflect the latest science, and that the scientific basis is rigorous and unbiased. In reviewing thousands of studies, the agency staff ultimately needs to establish a cutoff date, or be faced with the need for a continuous review. The current review was based on studies completed by mid-2009, but in the June 29, 2012, *Federal Register* notice EPA indicated that it

> is aware that a number of new scientific studies on the health effects of PM have been published since the mid-2009 cutoff date for inclusion in the Integrated Science Assessment. As in the last PM NAAQS review, the EPA intends to conduct a provisional review and assessment of any significant new studies published since the close of the Integrated Science Assessment, including studies that may be submitted during the public comment period on this proposed rule in order to ensure that, before making a final decision, the Administrator is fully aware of the new science that has developed since 2009. In this provisional assessment, the EPA will examine these new studies in light of the literature evaluated in the Integrated Science Assessment. This provisional assessment and a summary of the key conclusions will be placed in the rulemaking docket.[50]

Publication of the proposed PM NAAQS rule in the *Federal Register* on June 29, 2012,[51] started a nine-week public comment period that ran through August 31, 2012. EPA also held two public hearings for the proposal on July 17, 2012, in Philadelphia, PA, and July 19, 2012, in Sacramento, CA.[52] EPA's final determinations for revising the PM NAAQS published on January 15, 2013, were based on information provided in the two public hearings, the more than 230,000 written public comments received, and EPA's consideration of and analysis in response to this information. EPA also revised its Regulatory Impact Analysis (RIA),[53] in large part in response to comments received.

IMPLEMENTING THE REVISED $PM_{2.5}$ NAAQS

Promulgation of NAAQS sets in motion a process under which the states and EPA first identify geographic nonattainment areas, those areas failing to comply with the NAAQS based on monitoring and analysis of relevant air quality data.[54] The CAA is specific with regard to the timelines for determining areas in noncompliance, submission of plans for achieving (or maintaining) compliance, and when noncompliant areas must achieve the established or revised NAAQS.

Within three years of issuance of a NAAQS, states are required to submit "infrastructure" plans demonstrating that they have the basic air quality management components necessary to implement the NAAQS.[55] Following states' proposed and EPA's final designations of attainment and nonattainment areas, states (and tribes if they choose to do so) must submit their plans (State Implementation Plans, or SIPs) for how they will achieve and/or maintain attainment of the standards. These often include new or amended state regulations and new or modified permitting requirements.

If new, or revised, SIPs for attainment establish or revise a transportation-related emissions allowance ("budget"), or add or delete transportation control measures, they will trigger "conformity" determinations. Transportation conformity is required by the CAA, Section 176(c) (42 U.S.C. 7506(c)), to prohibit federal funding and approval for highway and transit projects unless they are consistent with ("conform to") the air quality goals established by a SIP, and will not cause new air quality violations, worsen existing violations, or delay timely attainment of the national ambient air quality standards.[56]

Areas designated nonattainment for the NAAQS also are subject to new source review (NSR) requirements. Enacted as part of the 1977 CAA Amendments and modified in the 1990 CAA Amendments, NSR is designed to ensure that newly constructed facilities, or substantially modified existing facilities, do not result in violation of applicable air quality standards. NSR provisions outline permitting requirements both for construction of new major pollution sources and for modifications to existing major pollution sources.[57] The specific NSR requirements for affected sources depend on whether the sources are subject to "Prevention of Significant Deterioration" (PSD) or nonattainment provisions.[58] As discussed earlier (see "EPA's January 2013 Final Changes to the PM NAAQS"), the January 2013 final PM NAAQS includes revisions to the PSD permitting program (rules) with respect to the revised PM NAAQS so as not to "unreasonably delay" pending permits and establish a "grandfather" provision for permit applications if a draft permit or preliminary determination has been issued for public comment by the date the revised PM NAAQS go into effect.

In addition to the CAA requirement for states to submit implementation plans, EPA acts to control NAAQS pollutants through national regulatory programs. These may be in the form of regulations of products and activities that might emit the pollutants (particularly fuels and combustion engines, such as automobiles and trucks) and in the form of emission standards for new stationary sources (e.g., utilities, refineries). Often these national regulations reflect aspects of state rules previously issued by various states. EPA anticipates that recent CAA rules, including rules to reduce pollution from power plants, clean diesel rules for vehicles, and rules to reduce pollution from stationary diesel engines, would help states meet the revised PM NAAQS.

Nonattainment Area Designation Process

The process of designating nonattainment areas is intended as a cooperative federal-state-tribal[59] process in which states and tribes provide initial designation recommendations to EPA for consideration. In Section 107(d)(1)(A) (42 U.S.C. 7407), the statute states that the governor of each state shall submit a list to EPA of all areas in the state, "designating as ... nonattainment, any area that does not meet (*or that contributes to ambient air quality in a nearby area that does not meet*) an air quality standard" (emphasis added). Areas are identified as "attainment/unclassified"[60] when they meet the standard or when the data are insufficient for determining compliance with the NAAQS.

Following state and tribal recommended designation submissions, the EPA Administrator has discretion to make modifications, including to the area boundaries. As required by statute (Section 107(d)1(B)(ii)), the agency must notify the states and tribes regarding any modifications, allowing them sufficient opportunity to demonstrate why a proposed modification is inappropriate, but the final determination rests with EPA.

Measuring and analyzing air quality to determine where NAAQS are not being met is a key step in determining an area's designation. Attainment or nonattainment designations are made primarily on the basis of three years of federally referenced monitoring data.[61] EPA began developing methods for monitoring fine particles at the time the $PM_{2.5}$ NAAQS were being finalized in 1997, and operation of the network of monitors for $PM_{2.5}$ was phased in from 1999 through 2000. The network of monitors and their locations have been modified over time. Most recently, in a separate action in conjunction with the October 2006 publication of the revised particulates NAAQS, EPA amended its national air quality monitoring requirements, including those for monitoring particle pollution.[62] The amended monitoring requirements were intended to help federal, state, and local air quality agencies by adopting improvements in monitoring technology. Additional modifications to the PM NAAQS monitoring network were included in the final January 2013 rule, as discussed earlier in this report.

In addition to air emission and air quality data, EPA considers a number of other relevant factors when designating nonattainment areas,[63] and recommends that states apply these factors in their determinations in conjunction with other technical guidance. Examples of these factors include population density and degree of urbanization (including commercial development), growth rates, traffic and commuting patterns, weather and transport patterns, and geography/topography. States and tribes may submit additional information on factors they believe are relevant for EPA to consider.

Nonattainment areas include those counties where pollutant concentrations exceed the standard as well as those that contribute to exceedance of the standard in adjoining counties. Entire metropolitan areas tend to be designated nonattainment, even if only one county in the area has readings worse than the standard. In addition to identifying whether monitored violations are occurring, states' or tribes' boundary recommendations for an area are to also show that violations are not occurring in those portions of the recommended area that have been excluded, and that they do not contain emission sources that contribute to the observed violations.

January 2013 Final Revised $PM_{2.5}$ Annual NAAQS: Potential Area Designations

The January 2013 final rule revising the $PM_{2.5}$ annual standard is expected to result in an increase in the number of areas (typically defined by counties or portions of counties) designated nonattainment. Similar to the revisions to the $PM_{2.5}$ daily (24-hour) standard in 2006, the January 2013 revised concentrations for the $PM_{2.5}$ annual standard are expected to affect primarily areas currently in nonattainment for the 2006 standards, but would also likely include a few counties that have not been previously designated as nonattainment. EPA would not require new nonattainment designations for PM_{10} primary NAAQS since the standards were not changed in the January 2013 final rule.

The effective date of final PM NAAQS revisions corresponds with the January 15, 2013, publication in the *Federal Register*. State and tribal area designation recommendations would be required under the CAA to be submitted to EPA by January 2014 (within one year of the effective date of the final rule). The CAA requires EPA to make its final area designations within one year of the state and tribal recommendations, projected to be January 2015. EPA is required to notify states and tribes of its intended modifications to their recommendations 120 days (projected to be August-September 2014) prior to promulgating final designations which are expected to become effective sometime in early 2015.

The actual area designations of nonattainment are more than two years away and will be based on more current monitoring data (likely 2011-2013) and several other factors. Based on anticipated reductions associated with several other existing national air pollution control regulations and programs (see discussion in "National Regulations" section), EPA predicted that seven counties in California would be the only areas unable to meet the new $PM_{2.5}$ primary standard by 2020.[64] Additionally, for illustrative purposes, EPA identified 66 counties with monitors that show concentrations of $PM_{2.5}$ that would exceed the revised limit of the primary annual standard of 12 $\mu g/m^3$ based on 2009-2011 air quality monitoring data.[65] According to EPA, 47 of these counties were determined nonattainment areas previously for $PM_{2.5}$ NAAQS based on earlier monitoring data available at the time and other factors considered.

The map in **Figure 1** below depicts the potential nonattainment areas (counties) for the revised $PM_{2.5}$ annual standards based solely on the 2009-2011 air quality monitoring. The areas are depicted in the map for illustration purposes as a rough approximation of the potential areas that may be designated nonattainment, as they do not take into account other factors generally considered in making final designation determinations. The specific counties based on the 2009- 2011 monitoring data are shown in **Appendix C**, which also shows the overlap of those nonattainment areas for the existing (2006) $PM_{2.5}$ annual and/or daily (24-hour) standards, and indicates those areas not previously designated nonattainment. An apt direct comparison of areas expected to be designated nonattainment for the final 2013 revised $PM_{2.5}$ standards based solely on monitoring data with *all* areas designated nonattainment (counties and portions of counties) for the prior (2006 and 1997) PM NAAQS[66] is not practical, since some counties may have been determined attainment or nonattainment at that time based on consideration of factors in addition to relevant monitoring. Overlaying those counties with monitors identified by EPA based only on 2009-2011 monitoring data provides some indication of potential areas that may be considered for nonattainment designations. EPA expects data from future monitoring, 2011-2013, will possibly show continued decline in levels of PM and their precursors, resulting in fewer nonattainment areas than shown by the 66 counties approximated.

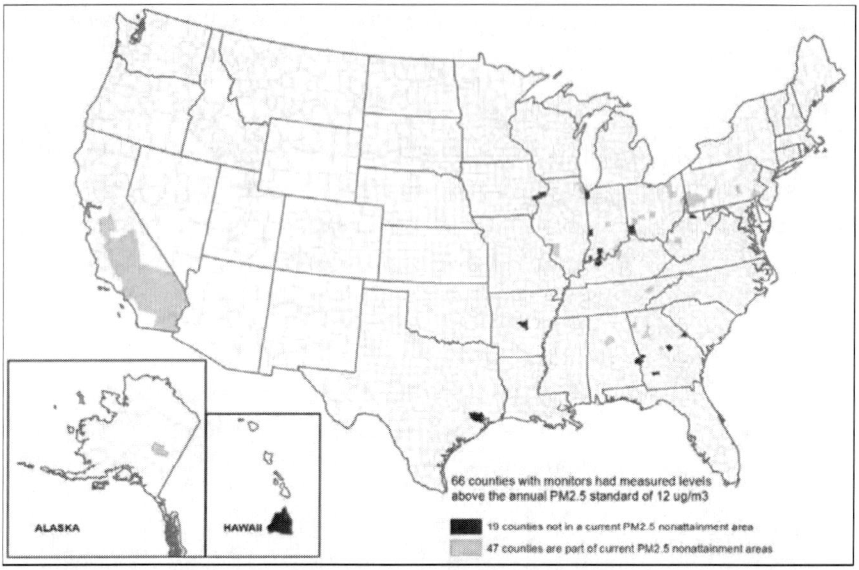

Source: Source: Created by CRS using data from the U.S. Environmental Protection Agency (EPA)— http://www.epa.gov/airquality/particlepollution/2012/ 20092011table.pdf. Base Map: Esri and U.S. Census. Projection: Lambert Conformal Conic. US_Counties_ FineParticle_juzel.mxd. February 4, 2013. EPA maps and supporting documents regarding the January 2013 $PM_{2.5}$ NAAQS revisions are available on EPA's website *Particulate Matter (PM): Regulatory Actions*, http://www.epa.gov/pm/ actions.html.

Notes: Forty-seven of the projected counties shown on the map were in nonattainment areas for $PM_{2.5}$ NAAQS based on a comparison with areas designated nonattainment previously by EPA for the $PM_{2.5}$ 24-hour standard (35 $\mu g/m^3$) as promulgated in 2006 and/or the annual standard (15 $\mu g/m^3$) that was retained in 2006 at the 1997 level. Specific counties and their comparisons are shown in the table in **Appendix C**. The designations are presented for illustrative purposes only. EPA will not designate areas as nonattainment for the revised $PM_{2.5}$ NAAQS based on 2009-2011 air monitoring data. Designations will most likely be based on 2011-2013 air monitoring data that the agency anticipates will indicate comparatively improved air quality.

Figure 1. Counties Not Meeting the January 2013 Revised Primary Annual $PM_{2.5}$ NAAQS Based on 2009-2011 Air Monitoring Data. (revised annual standard of 12 $\mu g/m^3$).

The 2006 revisions to the PM NAAQS tightening the 24-hour standard, which are currently being implemented, primarily affected urban areas. EPA published its final designations of 31 areas in 18 states, comprising 120 counties (89 counties and portions of 31 additional counties) for nonattainment of the revised 2006 *24-hour* $PM_{2.5}$ standard, on November 13, 2009.[67] Based on the 2009-2011 data, 28 of the 120 counties designated nonattainment for the 2006 PM 24-hour standard would be in nonattainment for the January 2013 annual standard. The designations, based on 2006 through 2008 air quality monitoring data, included a few counties that were designated nonattainment for $PM_{2.5}$ for the first time, but the majority of the counties identified overlapped with EPA's final nonattainment designations for the 1997 $PM_{2.5}$ NAAQS.[68] It is important to note that most of the 1997 $PM_{2.5}$ nonattainment areas were *only* exceeding the annual standard; thus, tightening the 24-hour standard resulted in an increased number of areas being designated nonattainment based on exceedances of both the 24-hour *and* the annual standard. The majority of the roughly 3,000 counties throughout the United States (including tribal lands) were designated attainment/unclassifiable, and are not required to impose additional emission control measures to reduce $PM_{2.5}$.

State Implementation Plans (SIPs)

The CAA requires that within three years of issuance of a NAAQS, states are required to submit "infrastructure" plans demonstrating that they have the basic air quality management components necessary to implement the NAAQS.[69] Areas designated attainment/unclassifiable will not have to take steps to improve air quality, but under the statute they must take steps to prevent air quality from deteriorating to unhealthy levels ("maintenance plans"). For those areas eventually designated nonattainment, state, local, and tribal governments must outline detailed control requirements in plans demonstrating how they will meet the revised primary annual $PM_{2.5}$ NAAQS.

These plans, defined as state implementation plans and referred to as SIPs (TIPs for tribal implementation plans), must be submitted to EPA three years after the effective date of the agency's final designations.[70] EPA projects final area designations will be effective early 2015 for the January 2013 PM NAAQS revisions, thus SIPs and TIPs would be required by early 2018. If states fail to develop an adequate implementation plan, EPA can impose one. Under the CAA, states are required to meet any established or revised $PM_{2.5}$

standard “as expeditiously as practicable,” but no later than five years from the effective date of designation—December 2020 according to EPA’s timeline—unless an extension (up to five additional years) allowed under the CAA is granted.[71] Changes in pollution control measures by affected sources (e.g., industrial operations, power plants, etc.) would not be required as a result of NAAQS revisions until after the area designations (see discussion in previous sections) are finalized or after the SIPs are finalized, depending on the specific circumstances.

National Regulations

EPA anticipates that in many cases, stationary and mobile source controls and additional reductions currently being adopted to attain the 2006 $PM_{2.5}$ standards in conjunction with expected emission reductions from implementing national regulations and strategies will help states meet the proposed standards. These national actions EPA referenced include the

- Cross-State Air Pollution Rule (CSAPR);[72]
- Mercury and Air Toxics Standards (MATS);[73]
- Light-Duty Vehicle Tier 2 Rule;[74]
- Heavy Duty Diesel Rule;[75]
- Clean Air Nonroad Diesel Rule;[76]
- Regional Haze Regulations and Guidelines for Best Available Retrofit Technology Determinations;[77]
- NOx Emission Standard for New Commercial Aircraft Engines;[78]
- Emissions Standards for Locomotives and Marine Compression-Ignition Engines;[79]
- Control of Emissions from Nonroad Spark Ignition Engines and Equipment;[80]
- Category 3 Oceangoing Vessels;[81]
- Reciprocating Internal Combustion Engines (RICE) National Emissions Standards for Hazardous Air Pollutants (NESHAPS);[82] and
- New Source Performance Standards and Emissions Guidelines for Hospital/Medical/Infectious Waste Incinerators Final Rule Amendments.[83]

Stakeholders and some Members of Congress are skeptical about EPA’s expectations with respect to the corollary benefits associated with some of these regulations, and raise concerns about pending efforts to delay some of the more recent programs and historical delays of others. Of particular concern are the Cross-State Air Pollution Rule ("Cross-State Rule," or CSAPR),[84] which was to have gone into effect in 2012 but was stayed in December 2011, then vacated on August 21, 2012, by the D.C. Circuit Court of Appeals,[85] and the Mercury and Air Toxics Standards (MATS), which EPA itself has stayed with regard to new plants, pending reconsideration. On October 5, 2012, the U.S. Department of Justice filed a petition[86] seeking en banc rehearing of the D.C. Circuit's August 21, 2012, decision regarding the CSAPR. The D.C. Circuit denied requests for both a panel and the en banc rehearing on January 24, 2013.[87] To date, EPA has not made a decision regarding its response to the Court's denial for a rehearing. Other rules remanded or reconsidered include the hazardous air pollutant ("MACT") standards for boilers and cement kilns. EPA has delayed implementation of the boiler MACT rules for more than a year and a half while considering changes to the requirements. The agency has also extended the compliance deadline for the cement kiln MACT by two years.

Potential Impacts of More Stringent PM Standards

The impacts of the revising PM NAAQS can be both potentially far-reaching and indirect. As discussed earlier in this report, the NAAQS by itself does not compel any specific direct pollution control measures. Rather it starts a process that could result in significant required investments by emitting sources in control measures. In addition to these costs, the eventual result is projected by EPA to be potentially significant health benefits. Estimates of health and welfare risk reductions and costs associated with control strategies for areas potentially not in compliance provide some insights into potential impacts of the June 2012 proposed and January 2013 final revisions to the PM NAAQS.

The Clean Air Act requires that NAAQS be set solely on the basis of public health and welfare protection, while costs and feasibility are generally taken into account in implementation of the NAAQS (a process that is primarily a state responsibility). As discussed previously, in setting and revising the NAAQS, the CAA directs the EPA Administrator to protect public health *with an adequate margin of safety*. This language has been interpreted, both by the agency and by the courts, as requiring standards be based on a review of the health impacts, without consideration of the costs, technological feasibility, or other non-health criteria.[88]

Nevertheless, coinciding with the PM NAAQS final rule released on December 14, 2012, and proposed rule in the June 29, 2012, *Federal Register*, EPA released regulatory impact analyses (RIA)[89] assessing the costs and benefits of setting the standard at the proposed and other alternative levels, to meet its obligations under Executive Order 12866 and in compliance with guidance from the White House Office of Management and Budget.[90] EPA emphasized that the RIA is for informational purposes and that decisions regarding revisions to the PM NAAQS are not based on consideration of the analyses in the RIA in any way. In addition, the expected costs are more difficult to predict than for many other regulations because the ultimate pollution control requirements, which are the primary costs, will depend on a variety of factors, such as state regulatory decisions and the results of monitoring and modeling analysis of designated areas that are not fully knowable at this time.

In part in response to comments received and considered following the June 2012 proposal, EPA revised its RIA for the final rule.[91] **Table 2** below presents a range of EPA's estimated economic costs, monetized benefits, and net benefits (subtracting total costs from the monetized benefits) associated with achieving the revised $PM_{2.5}$ standards in the final rule published in January 2013, and other alternatives considered as presented in EPA's revised RIA.

As shown in **Table 2**, EPA estimated that the monetized benefits associated with the January 2013 final revised $PM_{2.5}$ annual standard of 12 $\mu g/m^3$ would range $4.0 billion to $9.1 billion per year in 2020 (2010 $), compared to annual costs ranging from $53.0 million to $350.0 million. EPA also noted that a full accounting of benefits would include additional environmental and societal benefits that were not quantified in the analysis. The basis for the benefits calculations[92] is health and welfare impacts attributable to reductions in ambient concentration of $PM_{2.5}$ resulting from a reasonable, but "speculative," array of known state implementation emission control strategies selected by EPA for purposes of analysis. The analysis does not model the specific actions that each state will undertake or emerging technologies in implementing the alternative $PM_{2.5}$ NAAQS. EPA notes that reductions in annual premature deaths represent a substantial proportion of total monetized benefits (over 90%).[93]

Table 2. EPA's Estimated Total Monetized Benefits, Costs, and Net Benefits of Attaining Alternative $PM_{2.5}$ NAAQS as in 2020 for the January 2013 Final Rule (2010 $ in millions)

Final and Alternative Annual Standard (μg/m^3)	Estimated Monetized Benefits[a]		Estimated Total Costs[b]	Estimated Net Benefits	
	Discount Rate[c]				
	3%	7%	7%	3%	7%
13	$1,300 to $2,900		$11 to $100	$1,200 to $2,900	$1,100 to $2,600
12	$4,000 to $9,100	$3,600 to $8,200	$53 to $350	$3,700 to $9,000	$3,300 to $8,100
11	$13,000 to $29,000	$12,000 to $26,000	$320 to $1,700	$11,000 to $29,000	$10,000 to $26,000

Source: Adapted from Environmental Protection Agency's "U.S. EPA, "Regulatory Impact Analysis for the Final Revisions to the National Ambient Air Quality Standards for Particulate Matter," December 2012, Table ES-2, p. ES-15, http://www.epa.gov/pm/2012/finalria.pdf. Estimates and results are as reported by EPA and have been rounded after calculation.

Note: Results are rounded to two significant digits after calculation for presentation and computation as reported by EPA. Estimates (costs and benefits) reflect full attainment in 2020, which includes implementation of several national programs and are incremental to compliance with the 2006 $PM_{2.5}$ NAAQS. The discount rates are as recommended in EPA's *Guidelines for Preparing Economic Analyses (2000)* and OMB Circular A-4 (2003).

a. The reduction in premature deaths each year accounts for over 90% of total monetized benefits. Mortality risk evaluation assumes discounting over the Science Advisory Board-recommended 20-year segmented lag structure. Not all possible benefits or "disbenefits" are quantified and monetized in this analysis. Data limitations prevented EPA from quantifying these endpoints, and as such, these benefits are inherently more uncertain than those benefits that EPA was able to quantify.

b. The two cost estimates do not represent lower- and upper-bound estimates, but represent estimates generated by two different methodologies. The lower estimate is generated using the fixed-cost methodology, which assumes that technological change and innovation will result in the availability of additional controls by 2020 that are similar in cost to the higher end of the cost range for current, known controls. The higher estimate is generated using the hybrid methodology, which assumes that while additional controls may become available by 2020, they become available at an increasing cost, and the increasing cost varies by geographic area and by degree of difficulty associated with obtaining the needed emissions reductions.

c. Due to data limitations, EPA was unable to discount compliance costs for all sectors at the 3% discount rate. Consequently, the net benefit calculations at 3% were computed by subtracting the costs at the 7% rate from the monetized benefits with the 3% rate.

EPA estimated total costs under partial and full attainment of several alternative PM standards.[94] The engineering costs generally include the costs of purchasing, installing, and operating the referenced control technologies. The technologies and control strategies selected for analysis are illustrative of one way in which nonattainment areas could meet a revised standard. EPA anticipates that in actual SIPS, state and local governments will consider programs that are best suited for local conditions as there are various options for potential control programs that would bring areas into attainment with alternative standards. EPA includes a detailed discussion of the limitations and uncertainties associated with the benefits assumptions and analyses.[95]

While recognizing the need to adequately protect against potential health concerns associated with PM, some Members and stakeholders are apprehensive that EPA has underestimated potential costs and are concerned with the potential monetary consequences associated given the current economic environment. In particular, some stakeholders question the validity of EPA's reliance on the associated impacts of other national regulations in reducing the potential burdens. Critics are concerned that this results in underestimating the number of areas (counties) likely to be affected in terms of their ability to attain the proposed alternative PM NAAQS and the expected associated costs of necessary measures that will be required in the form of SIPs.

REACTION TO THE REVISED PM NAAQS

Prior to EPA's June 2012 proposed rule to revise the PM NAAQS, stakeholders were providing evidence and arguments in letters, press releases, at public hearings and other forums for their preferred recommendations, and EPA received numerous comments during various stages of development of the criteria and policy documents. In general, business and industry opposed more stringent standards particularly in light of the current national and global economic environment; and public health and environmental advocacy groups advocated support for more stringent standards based on the continuing evidence of health effects from ongoing scientific research. As mentioned earlier, several states petitioned EPA, and subsequently filed suit in the D.C. Circuit Court urging timely completion of EPA's review of the PM NAAQS in response to the February 2009 remand. Other state air quality regulators recognized the need to ensure adequate health protection from PM, but expressed concerns about the impacts of more stringent PM NAAQS on already strained state budgets.

Proponents of more stringent standards generally stress that

- the $PM_{2.5}$ standards should be at least as stringent as the more stringent combined daily and annual levels recommended in the 2006 EPA staff paper, and those recommended by the CASAC;
- scientific evidence of adverse health effects is more compelling than when the standards were revised in 2006;
- more stringent standards ensure continued progress toward protection of public health with an adequate margin of safety as required by the CAA; and
- welfare effects, particularly visibility, should be enhanced.

Critics of more stringent PM NAAQS stress that

- more stringent (and in some cases the existing) standards are not justified by the scientific evidence; the proposal does not take into account studies completed since the 2009 cutoff;
- requiring the same level of stringency for all fine particles without distinguishing sources is unfounded;
- costs and adverse impacts on regions and sectors of the economy are excessive;
- EPA has potentially overstated the expected benefits and underestimated expected costs;
- revising the standards could impede implementation of the existing (2006) PM NAAQS and the process of bringing areas into compliance, given the current status of this process;
- the benefits (and costs) associated with implementation of the 2006 PM NAAQS, as well as compliance with other relatively recent EPA air quality regulations that are being implemented, have not yet been realized; and

- revisions to PM NAAQS are unnecessary as shown by EPA's trends data that annual and 24-hour measured PM national concentrations have declined 24% and 28% respectively from 2001 to 2010.

CONGRESSIONAL ACTIVITY

Not long after EPA's release of its PM NAAQS proposal, the House Committee on Energy and Commerce Subcommittee on Energy and Power held a hearing on June 28, 2012,[96] on the potential impacts of tightening the $PM_{2.5}$ NAAQS. The focus of the debate was the regulatory costs and burdens associated with the implementation of the revised standards, and potential impacts on economic growth, employment and consumers. Just prior to EPA's release of the June 2012 proposal, several Members urged the Administrator to include retaining the current (as of 2006) $PM_{2.5}$ standard as an option for consideration in the agency's proposal.[97] In November 2012, some Members[98] urged EPA to consider delaying the final rule, while conversely, others, along with some state attorneys general,[99] supported timely completion of the agency's review. As mentioned earlier in this report, also in November 2012, some Members recommended EPA reconsider its calculations of costs and benefits supporting the proposed rule. Also, although the January 15, 2013, final rule did not modify the standards for inhalable "coarse" particles larger than 2.5 but smaller than 10 microns (PM_{10}), nor were modifications proposed in June 2012, some Members maintained a particular interest in EPA's consideration of the PM_{10} standards.

During the second session of the 111th and during the first session of the 112th Congress, some Members raised concerns in letters to the EPA Administrator and during oversight hearings[100] about EPA's staff draft reports and CASAC recommendations regarding changes to the PM NAAQS leading up to the June 2012 proposal. Some Members expressed their concerns of potential impacts that the options for changing PM NAAQS standards could have on industry and on agricultural operations. In letters to the EPA Administrator, several Members also communicated their particular concerns with the agency's consideration of stricter standards for coarse particulates (PM_{10}), including apprehensions of how changes may affect the agricultural community.[101] Additionally, during the 112th Congress, the House-passed Farm

Dust Regulation Prevention Act of 2011 (H.R. 1633) would have prohibited EPA from proposing, finalizing, implementing, or enforcing any regulation revising primary or secondary NAAQS applicable to PM "... with an aerodynamic diameter greater than 2.5 micrometers ..." for one year. Further, the House-passed bill would have amended the CAA to exempt "nuisance dust" from the act and would have excluded nuisance dust from references in the act to particulate matter "... except with respect to geographic areas where such dust is not regulated under state, tribal, or local law.... " Nuisance dust was defined in the bill as particulate matter that

> (1) is generated primarily from natural sources, unpaved roads, agricultural activities, earth moving, or other activities typically conducted in rural areas; (2) consists primarily of soil, other natural or biological materials, windblown dust, or some combination thereof; (3) is not emitted directly into the ambient air from combustion, such as exhaust from combustion engines and emissions from stationary combustion processes; (4) is not comprised of residuals from the combustion of coal; and (5) does not include radioactive particulate matter produced from uranium mining or processing.

A general provision included in FY2012 House-reported EPA appropriations language (H.R. 2584, Title IV, and §454)[102] would have restricted the use of FY2012 appropriations "to modify the national primary ambient air quality standard or the national secondary ambient air quality standard applicable to coarse particulate matter (generally referred to as "PM_{10}")."[103] No comparable provision was retained in the Consolidated Appropriations Act, 2012 (P.L. 112-74), enacted December 23, 2011, which ultimately included EPA's FY2012 appropriation.

NAAQS decisions have often been a source of significant concern to many in Congress. The evolution and development of the PM (and ozone) NAAQS, in particular, have been the subject of extensive oversight. For example, following promulgations of the 1997 NAAQS Congress held 28 days of hearings on the EPA rule. Congress enacted legislation specifying deadlines for implementation of the 1997 standard, funding for monitoring and research of potential health effects, and the coordination of the PM (and ozone) standard with other air quality regulations. During the 109th Congress, hearings were held regarding implementation and review of the PM NAAQS leading up to promulgations of the 2006 PM NAAQS.[104]

Because of the potential impacts PM NAAQS could have on both public health and the economy, EPA's final rule published on January 15, 2013, modifying these standards has generated mixed reactions from some Members, and the issue will likely be of continued interest in the 113th Congress.

CONCLUSION

EPA's changes to the PM NAAQS in its final rule published on January 15, 2013, following completion of its statutorily required review, have continued to garner attention and conflicting concerns among a diverse array of stakeholders, and in Congress. As evidenced by the history of the PM NAAQS, the level of scrutiny and oversight could increase in the coming months. Because both the health and economic consequences of particulate matter standards are potentially significant, the PM NAAQS are likely to remain a prominent issue in the 113th Congress.

EPA asserts that its review and analyses of scientific evidence showed that revising the PM NAAQS could potentially result in fewer adverse health effects for the general population and particularly sensitive populations such as children, asthmatics, and the elderly, as well as improved welfare effects. Nonetheless, concerns remain with regard to the potential associated costs. In its assessment of the impacts of revising the PM NAAQS, EPA expects relatively few additional areas (counties) will be in nonattainment and require more stringent pollution controls to achieve compliance. Industry, some Members, and some state representatives anticipate that the January 2013 revised PM NAAQS could result in more areas than anticipated by EPA being classified as nonattainment and needing to implement new controls on particulate matter. Further, they are concerned that stricter standards may mean more costs for the transportation and industrial sectors, including utilities, refineries, and the trucking industry, affected by particulate matter controls. Others stress that related ongoing control efforts from prior and recently promulgated actions are expected to reduce the potential number of nonattainment areas, or at least facilitate compliance.

EPA's review and establishment of the 1997 PM NAAQS was the subject of litigation and challenges, including a Supreme Court decision in 2001.[105] EPA's 1997 promulgation of standards for both coarse and fine particulate matter prompted critics to charge EPA with over-regulation and spurred environmental groups to claim that EPA had not gone far enough. Not only was the science behind the PM NAAQS challenged, but EPA was also accused of unconstitutional behavior. More than 100 plaintiffs sued to overturn the standard. Although EPA's decision to issue the standards was upheld unanimously by the Supreme Court, for the most part, stakeholders on both sides of the issue continued to advocate their recommendations for more stringent and less stringent PM standard. Several states and industry, agriculture, business, and environmental and public health advocacy groups petitioned the U.S. Court of Appeals for the District of Columbia Circuit, challenging certain aspects of EPA's revisions of the PM NAAQS as promulgated December 2006. A February 24, 2009, decision by the D.C. Circuit granted the petitions in part, denying other challenges, and remanded the standards to EPA for further consideration. The court did not specifically vacate the 2006 PM NAAQS and implementation is currently underway.

EPA received considerable (more than 230,000 written) comments in response to the June 2012 proposal. Concerned stakeholders may return to the courts or initiate challenges in response to the final standards published on January 15, 2013, thus potentially furthering delays in designating nonattainment areas, and states' development and implementation of SIPs.

APPENDIX A. CHRONOLOGICAL SUMMARY OF KEY MILESTONES SUBSEQUENT TO THE JANUARY 2013 PM NAAQS FINAL RULE

As part of the D.C. Circuit's decision and a related Consent Agreement, EPA agreed to issue final revised PM NAAQS by December 14, 2012. The timeline presented in **Table A-1** below reflects the most recent projected milestone dates subsequent to the January 15, 2013, publication of the final rule revising the PM NAAQS. These milestones are driven primarily by statutory requirements under the CAA, and are based on milestones identified in the June 29, 2012, *Federal Register* and EPA fact sheets accompanying the agency's proposed and final regulatory actions. The CAA does not specify a timeframe with regard to when states must meet secondary PM standards; relevant milestones are determined by EPA and states through the implementation planning process.

Table A-1. Milestone Chronology for Actions Subsequent to the January 2013 Final Revisions to the PM NAAQS

Actual and Projected Date	January 2013 Revised PM NAAQS Milestones
December 14, 2012, Final Rule Released (F.R. published on January 15, 2013)	The EPA Administrator signed the final rule on December 14, 2012, as per the D.C. Circuit June 2012 and as agreed to under a Consent Decree. The final rule was published in the F.R. on January 15, 2013.
January 2014 Proposal of Area Designations (required by CAA within one year after publication of PM NAAQS final rule)	State-tribal area designation recommendations (based on 2010-2012 monitoring data).
August-September 2014 EPA Response	EPA notifies states and tribes regarding modifications to their recommendations.
January 2015 Final Area Designations (required one year after states and tribes make recommendations)	EPA promulgates final area designations; expected effective data early 2015.
No Date Available (pending)	EPA proposes PM2.5 implementation rule.
Early 2016 (one year after the final designation effective date of early 2015)	States with new transportation projects submit conformity determination within one year of the effective date of nonattainment designation.
Not Available	EPA promulgates final PM2.5 implementation rule.
Early 2018 (3 years after final area designations effective date of January 15, 2013)	States and tribes are to submit revised implementation plans (SIPs) to achieve PM2.5 compliance in nonattainment areas required three years after final designations.
Early 2020-2025 (5-10 years after final area designations effective date of January 15, 2013)	CAA NAAQS statutory compliance deadline that States must meet the health standards "as expeditiously as practicable" but not later than five years after designations. A state may request a possible extension to 2025, depending on the severity of an area's fine particle pollution problems and the availability of pollution controls.

Source: Prepared by CRS based on U.S. Environmental Protection Agency fact sheets, technical documents, guidance accompanying the EPA PM NAAQS final rule published on January 15, 2013, 77 *Federal Register* 38889- 39055, and the proposed rule, 77 *Federal Register* 38889-39055, June 29, 2012. See http://www.epa.gov/pm/ actions.html.

APPENDIX B. SUPPORTING EPA SCIENTIFIC AND POLICY DOCUMENTS, AND CASAC REVIEW

Table B-1. Chronological Listing of EPA Workshops, and Technical and Policy Documents in Support of the 2013 Revised PM NAAQS

Workshop/Draft or Final Document	Date
Integrated Science Assessment for Particulate Matter: Call for Information	June 2007
Workshop to Discuss Policy-Relevant Science to Inform EPA's Integrated Plan for the Review of the Primary PM NAAQS - Final Agenda	July 2007
Workshop to Discuss Policy-Relevant Science to Inform EPA's Integrated Plan for the Review of the Secondary PM NAAQS - Final Agenda	July 2007
PM NAAQS Integrated Review Plan - Draft	October 2007
PM NAAQS Integrated Review Plan - Final	March 2008
Notice of Workshop to Review Initial Draft Materials for the PM Integrated Science Assessment	May 2008
Integrated Science Assessment for Particulate Matter - First External Review Draft	December 2008
PM NAAQS: Scope and Methods Plan for Urban Visibility Impact Assessment	February 2009
PM NAAQS: Scope and Methods Plan for Health Risk and Exposure Assessment	February 2009
Integrated Science Assessment for Particulate Matter - Second External Review Draft	July 2009
Particulate Matter Urban-Focused Visibility Assessment—External Review Draft	September 2009
Risk Assessment to Support the Review of the PM Primary National Ambient Air Quality Standards - External Review Draft	September 2009
Review of Urban Visibility Public Preference Studies (Final Report)	September 2009
Urban-Focused Visibility Assessment Data File	November 2009
Corrections to Relative Humidity Values Used in the Draft UFVA, Corrected Graphics, Tables, and Availability of Detailed Data File for Current Conditions	November 2009
Integrated Science Assessment for PM (Final Report)	December 2009
Particulate Matter Urban-Focused Visibility Assessment - Second External Review Draft	January 2010
Statistical Analysis of Existing Urban Visibility Preference Studies	February 2010
Corrections to Relative Humidity Values Used in the Draft Urban-Focused Visibility Assessment, Availability of Data File Comparing Incorrect RH Data to Corrected RH Data for Atlanta and Birmingham	February 2010

Table B-1. (Continued)

Workshop/Draft or Final Document	Date
Quantitative Health Risk Assessment for Particulate Matter—Second External Review Draft	February 2010
Revision to Section 3.3.5 of the Second External Review Draft of the PM Urban Visibility Assessment	March 2010
Analyses of PM2.5 Data for the PM NAAQS Review, Hassett-Sipple	March 2010
Quantitative Health Risk Assessment for Particulate Matter - Final Report	June 2010
Quantitative Health Risk Assessment for Particular Matter - Air Quality Data Files (for hybrid rollback-based analyses)	June 2010
Quantitative Health Risk Assessment for Particular Matter - Air Quality Data Files (for proportional and locally focused rollback-based analyses)	June 2010
Corrected Urban-Focused Visibility Assessment Data File	July 2010
Particulate Matter Urban-Focused Visibility Assessment - Final Document	July 2010
PM10 and PM10-2.5 Air Quality Analyses, Schmidt and Jenkins	July 2010
Particulate Matter Air Quality Data Requested from Epidemiologic Study Authors	July 2010
SANDWICH-Related Correction to the UFVA Data File, as Used for the Final Document	July 2010
Explanation of Error in Table 4-3 of the Final UFVA	July 2010
PM2.5 Air Quality Analyses	July 2010
Assessment of the Use of Speciated PM2.5 Mass-Calculated Light Extinction as a Secondary PM NAAQS Indicator of Visibility	November 2010
Simplified Approaches for Calculation of Hourly PM2.5 Light Extinction Values From Hourly PM2.5 Mass and Relative Humidity Data and 24-hour PM2.5 Composition Data	November 2010
Supplemental analysis of PM10 Air Quality from Locations Evaluated by Zanobetti and Schwartz (2009)	February 2011
PM2.5 Air Quality Analyses - Update	April 2011
PM10 and PM10-2.5 Air Quality Analyses	April 2011
PM2.5 Distributional Statistical Analyses	April 2011
Assessment of PM2.5 FEMs Compared to Collocated FRMs	April 2011
Investigation of 1-hour PM2.5 Mass Concentration Data from EPA-Approved Continuous Federal Equivalent Method Analyzers	April 2011
Documentation of Measurement Uncertainty Estimates of Collocated Chemical Speciation Network and IMPROVE Data for Use in the Secondary PM2.5 Standard for Visibility	June 2012
Recommendations for Sampling Artifact Correction for PM2.5 Organic Carbon	June 2012
Technical Analyses to Support Surrogacy Policy for Proposed Secondary PM2.5 NAAQS under NSR/PSD Programs	June 2012

Source: Prepared by CRS based on U.S. Environmental Protection Agency fact sheets, list of technical documents available on its website Technology Transfer Network

(TNN) National Ambient Air Quality Standards (NAAQS): Particulate Matter (PM) Standards—Documents from Current Review at http://www.epa.gov/ttn/naaqs/standards/pm/s_pm_index.html, and 77 *Federal Register* 38889-39055, June 29, 2012.

Table B-2. Chronological Listing of CASAC Reviews and Consultations

Review/Consultation	Date
CASAC Particulate Matter Review Panel's Consultation on EPA's Draft Integrated Review Plan for the National Ambient Air Quality Standards for Particulate Matter - Teleconference	November 2007
CASAC Particulate Matter Review Panel's Consultation on EPA's Draft Integrated Review Plan for the National Ambient Air Quality Standards for Particulate Matter - Report	January 2008
Consultation on Ambient Air Monitoring Issues Related to the Coarse Particle Speciation by the Clean Air Scientific Advisory Committee (CASAC) Ambient Air Monitoring & Methods Subcommittee (AAMMS)	March 2009
Review of EPA's Integrated Science Assessment for Particulate Matter (First External Review Draft December 2008)	May 2009
Consultation on EPA's Particulate Matter National Ambient Air	May 2009
Quality Standards: Scope and Methods Plan for Health Risk and Exposure Assessment	
Consultation on EPA's Particulate Matter National Ambient Air Quality Standards: Scope and Methods Plan for Urban Visibility Impact Assessment	May 2009
Review of Integrated Science Assessment for Particulate Matter (Second External Review Draft, July 2009)	November 2009
Review of Particulate Matter Urban-Focused Visibility Assessment (External Review Draft, September 2009)	November 2009
Review of Risk Assessment to Support the Review of the Particulate Matter (PM) Primary National Ambient Air Quality Standards—External Review Draft (September 2009)	November 2009
CASAC Review of Particulate Matter Urban-Focused Visibility Assessment—Second External Review Draft (January 2010)	April 2010
CASAC Review of Quantitative Health Risk Assessment for Particulate Matter—Second External Review Draft (February 2010)	April 2010
Review of the White Paper on Particulate Matter (PM) Light Extinction Measurements	April 2010
CASAC Review of Policy Assessment for the Review of the PM NAAQS—First External Review Draft (March 2010)	May 2010
CASAC Review of Policy Assessment for the Review of the PM NAAQS—Second External Review Draft (June 2010)	September 2010

Source: Prepared by CRS based on U.S. Environmental Protection Agency fact sheets, list of CASAC documents available on EPA's websites "EPA Clean Air Scientific

Advisory Committee (CASAC) Final Reports by Topic" at http://yosemite.epa.gov/sab/sabproduct.nsf/WebReportsbyTopicCASAC!OpenView, and 77 *Federal Register* 38889-39055, June 29, 2012.

Appendix C. Comparison of Potential Nonattainment Areas for the January 2013 Final Revised $PM_{2.5}$ Annual Standard with the Final Designations for the 2006 and 1997 $PM_{2.5}$ NAAQS

Table C-1. Nonattainment Areas for the January 2013 24-Hour $PM_{2.5}$ NAAQS as Estimated Using 2009-2011 Data, Final Designations 2006 24-Hour $PM_{2.5}$ NAAQS October 8, 2009, and Final Designations for the 1997 $PM_{2.5}$ NAAQS Annual

Designation Areas	1997 $PM_{2.5}$ NAAQS	2006 $PM_{2.5}$ NAAQS	2013 $PM_{2.5}$ NAAQS
	EPA Final Designations	EPA Final Designations	Proxy Designations (based on 2009-2011 Data)
	Annual Standard (15 $\mu g/m^3$)	24-Hour Standard (35 $\mu g/m^3$ 98th)	Annual Standard (12 $\mu g/m^3$)
	Counties and Partial Counties (p)		
ALABAMA			
Birmingham, AL[a]	Jefferson	Jefferson	Jefferson
	Shelby	Shelby	
	Walker (p)	Walker (p)	
Chattanooga, AL-TN-GA	Jackson (p)		
UNDEFINED[b]			Russell
ALASKA			
Fairbanks, AK		Fairbanks N. Star (p)	Fairbanks N. Star
ARIZONA			
Nogales, AZ			
Pinal, CA		Pinal (p) (designated February 3, 2011)[c]	

Designation Areas	1997 $PM_{2.5}$ NAAQS	2006 $PM_{2.5}$ NAAQS	2013 $PM_{2.5}$ NAAQS
	EPA Final Designations	EPA Final Designations	Proxy Designations (based on 2009-2011 Data)
	Annual Standard (15 μg/m^3)	24-Hour Standard (35 μg/m^3 98th)	Annual Standard (12 μg/m^3)
	Counties and Partial Counties (p)		
ARKANSAS			
UNDEFINED[b]			Pulaski
CALIFORNIA			
Chico, CA		Butte (p)	
Imperial County, CA		Imperial (p)	Imperial
Los Angeles, CA	Los Angeles (p)	Los Angeles (p)	Los Angeles
	Orange	Orange	
	Riverside (p)	Riverside (p)	Riverside
	San Bernardino (p)	San Bernardino (p)	San Bernardino
Sacramento, CA		El Dorado (p)	
		Placer (p)	
		Sacramento	
		Solano (p)	
		Yolo (p)	
San Francisco Bay Area, CA		Alameda	
		Contra Costa	
		Marin	
		Napa	
		San Francisco	
		San Mateo	
		Santa Clara	
		Solano (p)	
		Sonoma (p)	
San Joaquin Valley, CA	Fresno	Fresno	Fresno
	Kern (p)	Kern (p)	Kern
	Kings	Kings	Kings
	Madera	Madera	
	Merced	Merced	Merced
	San Joaquin	San Joaquin	

Table C-1. (Continued)

Designation Areas	1997 $PM_{2.5}$ NAAQS	2006 $PM_{2.5}$ NAAQS	2013 $PM_{2.5}$ NAAQS
	EPA Final Designations	EPA Final Designations	Proxy Designations (based on 2009-2011 Data)
	Annual Standard (15 $\mu g/m^3$)	24-Hour Standard (35 $\mu g/m^3$ 98th)	Annual Standard (12 $\mu g/m^3$)
	Counties and Partial Counties (p)		
	Stanislaus	Stanislaus	Stanislaus
	Tulare	Tulare	Tulare
Yuba City-Marysville, CA		Sutter	
		Yuba (p)	
CONNECTICUT			
New York, NY-NJ-CT	Fairfield	Fairfield	
	New Haven	New Haven	
DELAWARE			
Philadelphia-Wilmington, PA-NJ-DE	New Castle	New Castle	
DISTRICT OF COLUMBIA			
Washington, DC-MD-VA	Entire District		
GEORGIA			
Atlanta, GA	Barrow		
	Bartow		
	Carroll		
	Cherokee		
	Clayton		Clayton
	Cobb		
	Coweta		
	De Kalb		
	Douglas		
	Fayette		
	Forsyth		
	Fulton		Fulton
	Gwinnett		
	Hall		
	Heard (p)		
	Henry		

Designation Areas	1997 $PM_{2.5}$ NAAQS	2006 $PM_{2.5}$ NAAQS	2013 $PM_{2.5}$ NAAQS
	EPA Final Designations	EPA Final Designations	Proxy Designations (based on 2009-2011 Data)
	Annual Standard (15 $\mu g/m^3$)	24-Hour Standard (35 $\mu g/m^3$ 98th)	Annual Standard (12 $\mu g/m^3$)
	Counties and Partial Counties (p)		
	Newton		
	Paulding		
	Putnam (p)		
	Rockdale		
	Spalding		
	Walton		
Chattanooga, AL TN-GA	Catoosa		
	Walker		
Macon, GA	Bibb		Bibb
	Monroe (p)		
Rome, GA	Floyd		Floyd
UNDEFINED[b]			Dougherty
			Muscogee
			Richmond
			Wikinson
HAWAII			
UNDEFINED[b]			Hawaii
IDAHO			
Logan, UT-ID		Franklin (p)	
Pinehurst, ID			
ILLINOIS			
Chicago-Gary-Lake County, IL-IN	Cook		Cook
	DuPage		
	Grundy (p)		
	Kane		
	Kendall (p)		
	Lake		
	McHenry		
	Will		
St. Louis, MO-IL	Madison		Madison
	Monroe		

Table C-1. (Continued)

Designation Areas	1997 $PM_{2.5}$ NAAQS	2006 $PM_{2.5}$ NAAQS	2013 $PM_{2.5}$ NAAQS
	EPA Final Designations	EPA Final Designations	Proxy Designations (based on 2009-2011 Data)
	Annual Standard (15 μg/m^3)	24-Hour Standard (35 μg/m^3 98th)	Annual Standard (12 μg/m^3)
	Counties and Partial Counties (p)		
	Randolph (p)		
	St. Clair		St. Clair
INDIANA			
Chicago-Gary-Lake County, IL-IN	Lake		
	Porter		
Cincinnati-Hamilton, OH-KY-IN	Dearborn (p)		
Evansville, IN	Dubois		Dubois
	Gibson (p)		
	Pike (p)		
	Spencer (p)		
	Vanderburgh		
	Warrick		
Indianapolis, IN	Hamilton		
	Hendricks		
	Johnson		
	Marion		Marion
	Morgan		
Lafayette-Frankfort, IN			
Louisville, KY-IN	Clark		Clark
	Floyd		Floyd
	Jefferson (p)		
Vincennes, IN			
UNDEFINED[b]			Lake
			Spencer
			Vanderbugh
			Vigo
IOWA			

Designation Areas	1997 $PM_{2.5}$ NAAQS	2006 $PM_{2.5}$ NAAQS	2013 $PM_{2.5}$ NAAQS
	EPA Final Designations	**EPA Final Designations**	**Proxy Designations (based on 2009-2011 Data)**
	Annual Standard (15 $\mu g/m^3$)	**24-Hour Standard (35 $\mu g/m^3$ 98th)**	**Annual Standard (12 $\mu g/m^3$)**
	Counties and Partial Counties (p)		
Davenport-Moline-Rock Island, IA-IL			Scott
Muscatine, IA			Muscatine
KENTUCKY			
Cincinnati-Hamilton, OH-KY-IN	Boone		
	Campbell		
	Kenton		
Huntington-Ashland, WV-KY-OH	Boyd		
	Lawrence (p)		
Louisville, KY-IN	Bullitt		Bullitt
	Jefferson		Jefferson
Paducah-Mayfield, KY-IL			
UNDEFINED[b]			Daviess
MARYLAND			
Baltimore, MD	Anne Arundel		
	Baltimore City		
	Baltimore		
	Carroll		
	Harford		
	Howard		
Washington, DC-MD-VA	Charles		
	Frederick		
	Montgomery		
	Prince George's		
Martinsburg, WV-Hagerstown, MD	Washington		
MICHIGAN			
Detroit-Ann Arbor, MI	Livingston	Livingston	
	Macomb	Macomb	
	Monroe	Monroe	

Table C-1. (Continued)

Designation Areas	1997 $PM_{2.5}$ NAAQS	2006 $PM_{2.5}$ NAAQS	2013 $PM_{2.5}$ NAAQS
	EPA Final Designations	EPA Final Designations	Proxy Designations (based on 2009-2011 Data)
	Annual Standard (15 μg/m^3)	24-Hour Standard (35 μg/m^3 98th)	Annual Standard (12 μg/m^3)
	Counties and Partial Counties (p)		
	Oakland	Oakland	
	St. Clair	St. Clair	
	Washtenaw	Washtenaw	
	Wayne	Wayne	
Grand Rapids, MI			
MISSISSIPPI			
MISSOURI			
St. Louis, MO-IL	Franklin		
	Jefferson		
	St. Charles		
	St. Louis		
	St. Louis City		St. Louis City
MONTANA			
Libby, MT	Lincoln (p)		
NEW JERSEY			
New York, NY-NJ-CT	Bergen		
	Essex		
	Hudson		
	Mercer		
	Middlesex		
	Monmouth		
	Morris		
	Passaic	Passaic	
	Somerset	Somerset	
	Union	Union	
Philadelphia-Wilmington, PA-NJ-DE	Burlington	Burlington	
	Camden	Camden	
	Gloucester	Gloucester	
NEW YORK			
New York, NY-NJ-CT	Bronx	Bronx	
	Kings	Kings	

Designation Areas	1997 $PM_{2.5}$ NAAQS	2006 $PM_{2.5}$ NAAQS	2013 $PM_{2.5}$ NAAQS
	EPA Final Designations	EPA Final Designations	Proxy Designations (based on 2009-2011 Data)
	Annual Standard (15 $\mu g/m^3$)	24-Hour Standard (35 $\mu g/m^3$ 98th)	Annual Standard (12 $\mu g/m^3$)
	Counties and Partial Counties (p)		
	Nassau	Nassau	
	New York	New York	
	Orange	Orange	
	Queens	Queens	
	Richmond	Richmond	
	Rockland	Rockland	
	Suffolk	Suffolk	
	Westchester	Westchester	
NORTH CAROLINA			
Hickory, NC	Catawba		
Greensboro-Winston Salem- High Point, NC	Davidson		
	Guilford		
UNDEFINED			
OHIO			
Canton-Massillon, OH	Stark	Stark	
Cincinnati-Hamilton, OH-KY-IN	Butler		Butler
	Clermont		
	Hamilton		Hamilton
	Warren		
Cleveland-Akron-Lorain, OH	Ashtabula (p)		
	Cuyahoga	Cuyahoga	Cuyahoga
	Lake	Lake	
	Lorain	Lorain	
	Medina	Medina	
	Portage	Portage	
	Summit	Summit	Summit
Columbus, OH	Coshocton (p)		
	Delaware		
	Fairfield		
	Franklin		Franklin

Table C-1. (Continued)

Designation Areas	1997 $PM_{2.5}$ NAAQS	2006 $PM_{2.5}$ NAAQS	2013 $PM_{2.5}$ NAAQS
	EPA Final Designations	EPA Final Designations	Proxy Designations (based on 2009-2011 Data)
	Annual Standard (15 µg/m^3)	24-Hour Standard (35 µg/m^3 98th)	Annual Standard (12 µg/m^3)
	Counties and Partial Counties (p)		
	Licking		
Dayton-Springfield, OH	Clark		Clark
	Greene		
	Montgomery		Montgo
Huntington-Ashland, WV-KY-OH	Adams (p)		
	Gallia (p)		
	Lawrence		
	Scioto		
Parkersburg-Marietta, WV-OH	Washington		
Steubenville- Weirton, OH-WV	Jefferson	Jefferson	Jefferson
Wheeling, WV-OH	Belmont		
Youngstown, OH			
OREGON			
Klamath Falls, OR		Klamath (p)	
Oakridge, OR		Lane (p)	
PENNSYLVANIA			
Allentown, PA		Lehigh	
		Northampton	Northampton
Harrisburg-Lebanon-Carlisle, PA		Cumberland	
		Dauphin	
		Lebanon	Dauphin
		York	
Johnstown, PA	Cambria	Cambria	Cambria
	Indiana (p)	Indiana (p)	
Lancaster, PA	Lancaster	Lancaster	
Liberty-Clairton, PA	Allegheny (p)	Allegheny (p)	Allegheny
Philadelphia-Wilmington, PA-NJ-DE	Bucks	Bucks	
	Chester	Chester	Chester

Designation Areas	1997 $PM_{2.5}$ NAAQS	2006 $PM_{2.5}$ NAAQS	2013 $PM_{2.5}$ NAAQS
	EPA Final Designations	**EPA Final Designations**	**Proxy Designations (based on 2009-2011 Data)**
	Annual Standard (15 $\mu g/m^3$)	**24-Hour Standard (35 $\mu g/m^3$ 98th)**	**Annual Standard (12 $\mu g/m^3$)**
	Counties and Partial Counties (p)		
	Delaware	Delaware	Delaware
	Montgomery	Montgomery	
	Philadelphia	Philadelphia	
Pittsburgh-Beaver Valley, PA	Allegheny (p)	Allegheny (p)	Allegheny
	Armstrong (p)	Armstrong (p)	
	Beaver	Beaver	Beaver
	Butler	Butler	
	Greene (p)	Greene (p)	
	Lawrence (p)	Lawrence (p)	
	Washington	Washington	Washington
	Westmoreland	Westmoreland	Westmoreland
Reading, PA	Berks		
York, PA	York		
TENNESSEE			
Chattanooga, AL-TN-GA	Hamilton		
Clarksville, TN-KY			
Knoxville-Sevierville-La Follette, TN	Anderson	Anderson	
	Blount	Blount	
	Knox	Knox	Knox
	Loudon	Loudon	
	Roane (p)	Roane (p)	
TEXAS			
UNDEFINED[b]			Harris
UTAH			
Logan, UT-ID		Cache (p)	
Provo, UTSalt		Utah (p)	
Lake City, UT		Box Elder (p)	
		Davis	
		Salt Lake	
		Tooele (p)	
		Weber (p)	

Table C-1. (Continued)

Designation Areas	1997 $PM_{2.5}$ NAAQS	2006 $PM_{2.5}$ NAAQS	2013 $PM_{2.5}$ NAAQS
	EPA Final Designations	EPA Final Designations	Proxy Designations (based on 2009-2011 Data)
	Annual Standard (15 μg/m^3)	24-Hour Standard (35 μg/m^3 98th)	Annual Standard (12 μg/m^3)
	Counties and Partial Counties (p)		
VIRGINIA			
Washington, DC-MD-VA	Alexandria City		
	Arlington		
	Fairfax City		
	Fairfax Co		
	Falls Church City		
	Loudoun		
	Manassas City		
	Manassas Park City		
	Prince William		
WASHINGTON			
Seattle-Tacoma, WA		Pierce (p)	
WEST VIRGINIA			
Charleston, WV	Kanawha	Kanawha	Kanawha
	Putnam	Putnam	
Huntington-Ashland, WV-KY-OH	Cabell		Cabell
	Mason (p)		
	Wayne		
Martinsburg, WV-Hagerstown, MD	Berkeley		
Morgantown, WV			
Parkersburg-Marietta, WV-OH	Pleasants (p)		
	Wood		Wood
Steubenville- Weirton, OH-WV	Brooke	Brooke	Brooke
	Hancock	Hancock	Hancock
Wheeling, WV-OH	Marshall		Marshall
	Ohio		Ohio
UNDEFINED[b]			Marion
WISCONSIN			
Green Bay, WI			
Madison-Baraboo, WI			

Designation Areas	1997 $PM_{2.5}$ NAAQS	2006 $PM_{2.5}$ NAAQS	2013 $PM_{2.5}$ NAAQS
	EPA Final Designations	**EPA Final Designations**	**Proxy Designations (based on 2009-2011 Data)**
	Annual Standard (15 μg/m^3)	**24-Hour Standard (35 μg/m^3 98th)**	**Annual Standard (12 μg/m^3)**
	Counties and Partial Counties (p)		
Milwaukee-Racine, WI		Milwaukee	
		Racine	
		Waukesha	
	20 states and DC	18 states	16 states
	38 areas	31 areas	NA
	204 counties	120 counties	66 countie
	173 whole counties	90 whole counties	NA
	31 partial counties	30 partial counties	NA

Source: Compiled by CRS using data from EPA Fact Sheets accompanying the January 15, 2013, final PM NAAQS rule, and EPA PM Designation's websites. In some designated areas, EPA included cities in the total count of whole and partial counties, with the exception of the District of Columbia.

a. In the September 20, 2010, *Federal Register*, EPA announced its determination that a three-county (Jefferson, Shelby, and portion of Walker) Alabama nonattainment area (Birmingham) has attaining data for the 2006 24-hour $PM_{2.5}$ NAAQS (75 *Federal Register* 57186, September 20, 2010). The clean air data determination was based on certified ambient air monitoring data showing the area monitored as in attainment for the 2006 24-hour $PM_{2.5}$ NAAQS based on 2007-2009 data.

b. The "designated areas" including one or more counties (or portions of counties) are as defined in the final designations for the 2006 $PM_{2.5}$. Those counties identified as potential nonattainment areas for the January 2013 revised standards that were not part of previously defined $PM_{2.5}$ NAAQS designated areas are characterized as "UNDEFINED" designation areas.

c. In a February 3, 2011 final notice, EPA published designations of three areas as "nonattainment" or "unclassifiable/attainment" for the 2006 24-$PM_{2.5}$ NAAQS that were deferred in the November 13, 2009, promulgated designations, 76 *Federal Register* 6056-6066, http://www.epa.gov/pmdesignations/ 2006standards/ documents/2011-01/FR-2011-01.pdf.

d. In the August 25, 2008, *Federal Register*, EPA announced its determination that a three-county (Harrisburg, Lebanon, Carlisle) Pennsylvania nonattainment area for the 1997 $PM_{2.5}$ NAAQS was in attainment (73 *Federal Register* 49949, August 25, 2008). The determination was based on certified ambient air monitoring data showing the area monitored as in attainment for the 1997 $PM_{2.5}$ NAAQS since the 2004-2006 monitoring period.

End Notes

1 The final rule and supporting documents are available on EPA's website *Particulate Matter (PM): Regulatory Actions*, http://www.epa.gov/pm/actions.html.

2 Sections 108-109 of the Clean Air Act (CAA) govern the establishment, review, and revisions of the NAAQS (42 U.S.C. 7408 and 7409).

3 Section 109(d)(1)) of the CAA.

4 *American Lung Ass'n v. EPA*, D.D.C., No. 1:12-cv-243, order issued June 6, 2012.

5 *American Lung Ass'n v. EPA*, D.D.C., No. 1:12-cv-243, order signed September 4, 2012. See also U.S. EPA, "Proposed Consent Decree," 77 *Federal Register* 38060, June 26, 2012, http://www.gpo.gov/fdsys/search/pagedetails.action?granuleId=2012-5603&packageId=FR-2012-06-26&acCode=FR, and *American Lung Ass'n v. EPA*, D.D.C., No. 1:12-cv-243, joint motion filed June 5, 2012.

6 U.S. EPA, National Ambient Air Quality Standards for Particulate Matter, Proposed Rule, 77 *Federal Register* 38889-39055, June 29, 2012. The proposal as signed by EPA Administrator Lisa P. Jackson on June 14, 2012, and supporting documents are available on EPA's website *Particulate Matter (PM): Regulatory Actions*, http://www.epa.gov/pm/actions.html.

7 U.S. EPA, Public Hearings for Proposed Rules—National Ambient Air Quality Standards for Particulate Matter, 77 *Federal Register* 39205, July 2, 2012.

8 For information regarding the CASAC PM review panel and its activities and reports, see http://yosemite.epa.gov/sab/sabpeople.nsf/WebCommittees/CASAC.

9 Section 109(d)(2) of the Clean Air Act.

10 The current review was initiated with EPA's June 2007 general call for information, U.S. EPA, "Integrated Science Assessment for Particulate Matter: Call for Information," 72 *Federal Register* 35462, June 28, 2007. See also EPA's *Policy Assessment for the Review of the Particulate Matter National Ambient Air Quality Standards*, pp. 1-10 through 1-12, U.S. EPA Office of Air Quality Planning and Standards, Health and Environmental Impacts Division, EPA 452/R-11-003, April 2011, http://www.epa.gov/ttnnaaqs/standards/pm/data/20110419pmpafinal.pdf.

11 See EPA's Fact Sheet, *Overview of EPA's Proposal to Revise the Air Quality Standards for Particle Pollution (Particulate Matter)*, http://www.epa.gov/pm/2012/fsoverview.pdf.

12 See CRS Report R41622, *Environmental Regulation and Agriculture*, coordinated by Megan Stubbs.

13 U.S. EPA, *Regulatory Impact Analysis for the Final Revisions to the National Ambient Air Quality Standards for Particulate Matter*, http://www.epa.gov/pm/2012/finalria.pdf. See also U.S. EPA, *Regulatory Impact Analysis for the Proposed Revisions to the National Ambient Air Quality Standards for Particulate Matter*, EPA 452/R-12-003, June 2012, http://www.epa.gov/ttn/ecas/regdata/RIAs/PMRIACombinedFile_Bookmarked.pdf. The RIA and supporting documents are available in the public docket, Docket No. EPA-HQ-OAR-2010-0955, http://www.regulations.gov/#!searchResults;rpp=25;po=0;s=EPA-HQ-OAR-2010-0955.

[14] For an example, see *Health Benefits of Alternative $PM_{2.5}$ Standards*, Donald McCubbin, Ph.D., prepared for the American Lung Association, Clean Air Task Force and Earthjustice, July 2011, http://earthjustice.org/sites/default/files/Health-Benefits-Alternative-PM2.5-Standards. pdf.

[15] See CRS Report R41561, *EPA Regulations: Too Much, Too Little, or On Track?*, by James E. McCarthy and Claudia Copeland.

[16] See November 21, 2012, letter from 47 Members of the House of Representatives to the U.S. EPA Administrator, http://latta.house.gov/uploadedfiles/2012_11_29_final_pm2_5_letter_signed_w_attchmt.pdf. Also see press release available on Representative Bob Latta's website at http://latta.house.gov/news/documentsingle.aspx?DocumentID= 314585.

[17] See December 6, 2012, letter from nine State Attorneys General to the Acting Administrator of the Office of Information and Regulatory Affairs, the White House Office of Management and Budget, http://www.eenews.net/ assets/2012/12/10/document_gw_02.pdf.

[18] 42 U.S.C. 7408(a)(1).

[19] The NAAQS is for NO_2; nitrogen gases that are ozone precursors are referred to as NOx.

[20] 42 U.S.C. 7408(a)(1).

[21] The use of public welfare in the CAA "includes, but is not limited to, effects on soils, water, crops, vegetation, manmade materials, animals, wildlife, weather, visibility, and climate, damage to and deterioration of property, and hazards to transportation, as well as effects on economic values and on personal comfort and well-being, whether caused by transformation, conversion, or combination with other air pollutants" (42 U.S.C. 7602(h)).

[22] Beginning in 1971, regulation and monitoring of particulate matter under the CAA focused primarily on total suspended particles (TSP) and, eventually in 1987, on coarse particles equal to or less than 10 micrometers in diameter (PM_{10}). EPA revised the particulates standards in 1997 to provide separate requirements for fine particulate matter ($PM_{2.5}$). See EPA's "Particulate Matter (PM) Standards—Table of Historical PM NAAQS" at http://www.epa.gov/ttn/ naaqs/standards/pm/s_pm_history.html.

[23] For a more detailed discussion regarding the petitions see section entitled "Petitions Challenging the 2006 PM NAAQS and the D.C. Circuit's February 29, 2009, Decision" in CRS Report RL34762, *The National Ambient Air Quality Standards (NAAQS) for Particulate Matter (PM): EPA's 2006 Revisions and Associated Issues*, by Robert Esworthy.

[24] *American Lung Ass'n v. EPA*, D.D.C., No. 1:12-cv-243, filed February 14, 2012.

[25] *States of New York, California, Connecticut, Delaware, Maryland, New Mexico, Oregon, Rhode Island, Vermont, and Washington, and Commonwealth of Massachusetts v. EPA*, D.S. N.Y., 12 CIV 1064, filed February 10, 2012, http://www.atg.state.vt.us/assets/files/NY%20v%20EPA%20Complaint%20(2-10-12).pdf.

[26] See footnote 4.

[27] See footnote 5.

[28] 62 *Federal Register* 38652-38896, July 18, 1997.

[29] PM_{10} NAAQS were promulgated in 1987, 52 *Federal Register* 24640, July 1, 1987.

[30] "The "form" of a standard defines the air quality statistic that is to be compared to the level of the standard in determining whether an area attains that standard." 77 *Federal Register* 38954, June 29, 2012.

[31] See footnote 1.

[32] Based on the findings in the EPA PM criteria document and staff paper, and the CASAC's concurrence, that the studies reviewed do not provide sufficient evidence regarding *long-term* exposure to warrant continuation of an annual standard. See 71 *Federal Register* 2653, *Section III. Rationale for Proposed Decision on Primary PM_{10} Standards,* January 17, 2006.

[33] Includes a measure of $PM_{2.5}$ mass, elements, ions, and carbon species. See EPA's laboratory standard operating procedures (SOPs) for $PM_{2.5}$ chemical speciation at http://www.epa.gov/ttnamti1/specsop.html.

[34] "The deciview scale is frequently used in the scientific and regulatory literature on visibility. This metric describes changes in uniform light extinction that can be perceived by a human observer. One deciview represents the minimal perceptible change in visibility to the human eye," 77 *Federal Register* 39043, June 29, 2012. A "deciview is a yardstick for measuring visibility: the higher the deciview level, the hazier the air appears," U.S. EPA, Fact Sheet: *Revised Air Quality Standards for Particle Pollution and Updates to the Air Quality Index (AQI)*, http://www.epa.gov/ pm/2012/decfsstandards.pdf.

[35] See U.S. EPA, "EPA's Regional Haze Program," http://www.epa.gov/visibility/program.html.

[36] See footnote 34.

[37] See EPA Fact Sheet: *EPA's Revised Air Quality Standards for Particle Pollution: Monitoring, Designations and Permitting Requirements*, http://www.epa.gov/pm/2012/decfsimp.pdf. See also EPA Fact Sheet: *EPA's Proposal to Update the Air Quality Standards for Particle Pollution: Monitoring, Designations and Permitting Requirements*, http://www.epa.gov/ airquality/particlepollution/2012/fsimp.pdf.

[38] EPA indicated that it is not increasing the size of the current $PM_{2.5}$ monitoring network of about 900 monitors, but anticipates that states will be able to relocate roughly 52 existing monitors to meet the near-roadway requirement; see previous footnote.

[39] U.S. EPA, Fact Sheet: *See EPA Fact Sheet: Revised Air Quality Standards for Particle Pollution and Updates to the Air Quality Index (AQI),* http://www.epa.gov/pm/2012/ decfsstandards.pdf. *See also EPA Fact Sheet: Summary of Proposed Improvements to the Air Quality Standards for Particle Pollution and Updates to the Air Quality Index (AQI),* http://www.epa.gov/pm/pdfs/PMNAAQSProposalSTANDARDSAQI61412FINALUPDAT ED.pdf.

[40] See footnote 37.

[41] For a detailed overview of the NAAQS process see CRS Report 97-722, *Air Quality Standards: The Decisionmaking Process.*

[42] For general information regarding the CASAC as well as the CASAC panel for the PM NAAQS review, see *EPA Clean Air Advisory Committee (CASAC)* website http://yosemite.epa.gov/sab/sabpeople.nsf/WebCommittees/CASAC.

[43] For EPA's most recent revisions to the CASAC review process, see the May 21, 2009, memorandum from Administrator Lisa P. Jackson to Dr. Jonathan Samet, CASAC Chair, and to Elizabeth Craig, Acting EPA Administrator for Air and Radon, and Lek Kadeli, Acting Administrator for Research and Development, http://yosemite.epa.gov/sab/ sabproduct.nsf/WebCASAC/NewNAAQSProcess?OpenDocument.

[44] The CAA directs the EPA Administrator to protect public health *with an adequate margin of safety*. This language has been interpreted, both by the agency and by the courts, as requiring standards based on a review of the health impacts, without consideration of the costs, technological feasibility, or other non-health criteria. Costs and feasibility are generally taken into account in NAAQS implementation (a process that is primarily a state responsibility).With regard to the non-relevance of cost considerations, see generally Whitman v. American Trucking Associations, 531 U.S. 457, 465-472, 475-76 (2001).

[45] U.S. EPA, "Integrated Science Assessment for Particulate Matter: Call for Information," 72 *Federal Register* 35462, June 28, 2007.

[46] U.S. EPA, *Policy Assessment for the Review of the Particulate Matter National Ambient Air Quality Standards*, U.S. EPA Office of Air Quality Planning and Standards, Health and Environmental Impacts Division, EPA 452/R-11-003, April 2011, http://www.epa.gov/ ttnnaaqs/standards/pm/data/20110419pmpafinal.pdf.

[47] Until discontinued by the CASAC Chairman in 2005, CASAC historically had signed off in the form of a "closure letter" *only* when the panel of members was convinced that each document accurately reflected the status of the science. The CASAC closure letter was an indication that the majority of the CASAC panel members had generally reached consensus that the criteria documents and the staff paper provided an adequate scientific basis for regulatory decision making. The discontinuance of the closure letter was the subject of

considerable debate, particularly within the science community. See CRS Report RL33807, *Air Quality Standards and Sound Science: What Role for CASAC?*, by James E. McCarthy.

[48] See 77 *Federal Register* 38900-38944, *Section III. Rationale for Proposed Decisions on Primary $PM_{2.5}$ Standards*, June 29, 2012.

[49] See 77 *Federal Register* 38944-38963, *Section IV. Rationale for Proposed Decisions on Primary PM_{10} Standards*, June 29, 2012.

[50] See 77 *Federal Register* 38899, *Section II. Background (B) Review of the Air Quality Criteria and Standards for PM (3) Current PM NAAQS Review*, June 29, 2012.

[51] U.S. EPA, National Ambient Air Quality Standards for Particulate Matter, Proposed Rule, 77 *Federal Register* 38889-39055, June 29, 2012. The proposal as signed by EPA Administrator Lisa P. Jackson on June 14, 2012 and supporting documents are available on EPA's website *Particulate Matter (PM): Regulatory Actions*, http://www.epa.gov/pm/actions.html.

[52] U.S. EPA, Public Hearings for Proposed Rules—National Ambient Air Quality Standards for Particulate Matter, 77 *Federal Register* 39205, July 2, 2012.

[53] For key components of the revised RIA see "Important Updates and Analytic Differences Between the PM NAAQS Proposal RIA and the Final RIA," Section ES.5 p. ES-23 of the December 2012 RIA, http://www.epa.gov/pm/2012/ finalria.pdf.

[54] For a general overview of the NAAQS designations process, see EPA's "Designations" website at http://www.epa.gov/air/urbanair/designations.html.

[55] Section 110(a)(2) of the Clean Air Act. For a general overview of the NAAQS implementation plans process, see EPA's "State Implementation Plan Overview" website at http://www.epa.gov/air/urbanair/sipstatus/overview.html.

[56] On March 14, 2012, EPA published a final rule restructuring sections of the conformity rule so that existing requirements apply to new or revised NAAQS and released associated implementation guidance July 2012. (U.S. EPA, Office of Transportation and Air Quality, *Guidance for Transportation Conformity Implementation in Multi- Jurisdictional Nonattainment and Maintenance Areas*, July 2012, http://www.epa.gov/otaq/stateresources/transconf/ regs/420b12046.pdf). For transportation conformity regulations see, U.S. EPA "State and Local Transportation Resources: Transportation Conformity" at http://www.epa.gov/otaq/stateresources/transconf/index.htm.

[57] For an overview, including statutory authority and regulations, see EPA's "New Source Review (NSR)" at http://www.epa.gov/air/nsr/.

[58] See Clean Air Act, Part D—Plan Requirements for Nonattainment Areas, sections 171-178, codified at 40 CFR 52.24(f)(10). Section 166 of the CAA authorizes EPA to establish regulations for PSD of any pollutant for which EPA has issued a national standard.

[59] Though not required, tribes have been encouraged to submit recommendations. The area designation requirements under the CAA (Section 107) are specific with respect to states, but not to tribes. EPA follows the same designation process for tribes per Sections 110(o) and 301(d) of the CAA and pursuant to the 1988 Tribal Authority Rule, which specifies that tribes shall be treated as states in selected cases (40 CFR Part 49). For information regarding tribes that have participated in the $PM_{2.5}$ designation recommendation process, see http://www.epa.gov/pmdesignations.

[60] Section 107(d)(1)(A)(iii) of the CAA provides that any area that EPA cannot designate on the basis of available information as meeting or not meeting the standards should be designated unclassifiable.

[61] A federally referenced monitor is one that has been accepted for use by EPA for comparison of the NAAQS by meeting the design specifications and certain precision and bias (performance) specifications (40 CFR Part 58).

[62] Revisions to Ambient Air Monitoring Regulations, final rule, 71 *Federal Register* 61235-61328, October 17, 2006. http://www.epa.gov/air/particlepollution/actions.html.

63 See Chapter 5 of the EPA Technical Support Document for December 17, 2004, final designations for the 1997 $PM_{2.5}$ NAAQS and April 2005 modifications, for explanations of these factors; available at http://www.epa.gov/ pmdesignations/1997standards/tech.htm.

64 See list of counties (http://www.epa.gov/pm/2012/2020table.pdfand) and map (http://www.epa.gov/pm/2012/ 2020map.pdf) depicting EPA's predictions for 2020, available on EPA's website *Particulate Matter (PM): Regulatory Actions*, http://www.epa.gov/pm/actions.html.

65 At the time of the June 2012 proposal, EPA had identified counties with monitors that showed concentrations of $PM_{2.5}$ that would exceed the proposed revised range of the primary annual standard of 12 $\mu g/m^3$ to 13 $\mu g/m^3$ based on 2008-2010 monitoring data.

66 For additional information, see CRS Report R40096, *2006 National Ambient Air Quality Standards (NAAQS) for Fine Particulate Matter (PM2.5): Designating Nonattainment Areas*, by Robert Esworthy.

67 74 *Federal Register* 58688-58781, November 13, 2009; see also "Area Designations for 2006 24-Hour Fine Particulate ($PM_{2.5}$) Standards—Regulatory Actions," http://www.epa.gov/pmdesignations/2006standards/regs.htm#4. Publication of a final area designation rule for the 2006 24-hour $PM_{2.5}$ NAAQS had been delayed as a result of the incoming Administration's review of the final rule, along with several other agency proposed and final actions introduced toward the end of the previous Administration. See footnote 66.

68 For detailed $PM_{2.5}$ state/county geographical designation recommendations by EPA and those from individual states and tribes, for the 1997 and for the 2006 $PM_{2.5}$ NAAQS, see http://www.epa.gov/pmdesignations.

69 Section 110(a)(2) of the Clean Air Act. For a general overview of the NAAQS implementation plans process, see EPA's "State Implementation Plan Overview" at http://www.epa.gov/air/urbanair/sipstatus/overview.html.

70 Section 172 of the Clean Air Act. See EPA's "State Implementation Plan Overview" at http://www.epa.gov/air/ urbanair/sipstatus/overview.html.

71 Under Section 172(a)(2)(A) of the CAA, EPA may grant an area an extension of the initial attainment date for one to five years (in no case later than 10 years after the designation date for the area). A state requesting an extension must submit an implementation plan (SIP) by the required deadline that includes, among other things, sufficient information demonstrating that attainment by the initial attainment date is "impracticable."

72 76 *Federal Register* 48208-48483, August 8, 2011. The CSAPR was intended to replace EPA's 2005 Clean Air Interstate Rule (CAIR, 70 *Federal Register* 25162, May 12, 2005), promulgated under the CAA, 42 U.S.C. 7401 et seq.; see http://www.epa.gov/crossstaterule/index.html. The CAIR had been remanded by the D.C. Circuit and to EPA in 2008 (North Carolina v. EPA 531 F.3d 896 (D.C. Cir. 2008)). For information regarding the CAIR rule see CRS Report RL34589, *Clean Air After the CAIR Decision: Multi-Pollutant Approaches to Controlling Powerplant Emissions*, by James E. McCarthy, Larry Parker, and Robert Meltz, and EPA "Clean Air Interstate Rule" at http://www.epa.gov/air/interstateairquality/#older.

73 77 *Federal Register* 9304-9513, February 16, 2012.

74 65 *Federal Register* 6822-6870, February 10, 2000.

75 65 *Federal Register* 59896-59978, October 6, 2000.

76 69 *Federal Register* 38958-39273, January 29, 2004.

77 70 *Federal Register* 39104-39172, July 6, 2005.

78 70 *Federal Register* 69644-69687, November 17, 2005.

79 73 *Federal Register* 37095-37144, republished June 30, 2008.

80 73 *Federal Register* 59034-59380, October 8, 2008.

81 75 *Federal Register* 22896-23065, April 30, 2010.

82 75 *Federal Register* 51570-51608, August 20, 2010; Proposed Amendments 77 *Federal Register* 33812-33857, June 7, 2012.

83 74 *Federal Register* 51415, October 6, 2009.

[84] See U.S. EPA, "Federal Implementation Plans: Interstate Transport of Fine Particulate Matter and Ozone and Correction of SIP Approvals," 76 *Federal Register* 48208-48483, August 8, 2011, http://www.gpo.gov/fdsys/pkg/FR- 2011-08-08/pdf/2011-17600.pdf. Explanatory and background material can be found on EPA's website at http://www.epa.gov/crossstaterule/ actions.html. See also footnote 72.

[85] *EME Homer City Generation, L.P. v. Environmental Protection Agency*, D.C. Cir., No. 11-1302, August 21, 2012, http://www.cadc.uscourts.gov/internet/opinions.nsf/ 19346B280C78405C85257A61004DC0E5/$file/11-1302- 1390314.pdf. See also U.S. EPA's website, "Cross-State Air Pollution Rule (CSAPR)," http://epa.gov/crossstaterule/ for this decision and other related documents.

[86] U.S. EPA, http://epa.gov/crossstaterule/pdfs/Rehearing_Petition_617874.pdf. For status of the petition see EPA website, "Cross-State Air Pollution Rule (CSAPR)," at http://epa.gov/crossstaterule/.

[87] *EME Homer City Generation, L.P. v. Environmental Protection Agency*, D.C. Cir., No. 11-1302, January 24, 2013, Orders. See U.S. EPA's website, "Cross-State Air Pollution Rule (CSAPR)," http://epa.gov/crossstaterule/.

[88] With regard to the non-relevance of cost considerations, see generally Whitman v. American Trucking Associations, 531 U.S. 457, 465-472, 475-76 (2001).

[89] U.S. EPA, "*Regulatory Impact Analysis for the Final Revisions to the National Ambient Air Quality Standards for Particulate Matter*," http://www.epa.gov/pm/2012/finalria.pdf, and U.S. EPA, "Regulatory Impact Analysis for the Proposed Revisions to the National Ambient Air Quality Standards for Particulate Matter," EPA-452/R-12-003 June 2012, available at http://www.epa.gov/ttn/ecas/ria.html.

[90] 58 *Federal Register* 51735, October 4, 1993. See the White House OMB website, *Regulatory Matters*, at http://www.whitehouse.gov/omb/regulatory_affairs/default.

[91] See footnote 53.

[92] See p. Section ES.2.2. beginning on p. ES-10, and discussion of health benefits in Chapter 5 beginning p. 5-1, and welfare benefits in Chapter 6 p. 6-1 of the EPA December 2012 RIA, footnote 89.

[93] U.S. EPA, p. ES-15 December 2012 RIA; see footnote 89.

[94] See discussion for engineering cost analysis in Chapter 7 beginning p. 7-1 (pdf p. 455) June 2012 RIA, footnote 89.

[95] See the Executive Summary in the RIA accompanying the January 2013 final rule: ES.4 Caveats and Limitations, beginning on p. ES-21.

[96] House Committee on Energy and Commerce Subcommittee on Energy and Power June 28, 2012 hearing entitled, "The American Energy Initiative: A Focus on the New Proposal to Tighten National Standards for Fine Particulate Matter," http://energycommerce.house. gov/hearing/american-energy-initiative-focus-new-proposal-tighten-national- standards-fine-particulate.

[97] See joint letter from Representatives Fred Upton, Chairman, Committee on Energy and Commerce, Ed Whitfield, Chairman, Subcommittee on Energy and Power, and Joe Barton, Chairman Emeritus, June 6, 2012, http://energycommerce.house.gov/letter/letter-epa-regarding-national-ambient-air-quality-standards.

[98] See November 21, 2012, letter from 47 Members of the House of Representatives to the U.S. EPA Administrator, http://latta.house.gov/uploadedfiles/2012_11_29_final_pm2_5_letter_ signed_w_attchmt.pdf. Also see press release available on Representative Bob Latta's website at http://latta.house.gov/news/documentsingle.aspx?DocumentID= 314585.

[99] See December 6, 2012, letter from nine State Attorneys General to the Acting Administrator of the Office of Information and Regulatory Affairs, the White House Office of Management and Budget, http://www.eenews.net/ assets/2012/12/10/document_gw_02.pdf.

[100] For example, U.S. Congress, Senate Committee on Agriculture, Nutrition, and Forestry, *Oversight Hearing to Examine the Impact of EPA Regulation on Agriculture*, 111th Cong., 2nd sess., September 23, 2010; and U.S. Congress, House Committee on Agriculture, *Public*

Hearing to Review the Impact of EPA Regulation on Agriculture, 112th Cong., 1st sess., March 10, 2011.

[101] Examples of letters to EPA Administrator Lisa Jackson include, but are not limited to, a joint letter from 99 House Members, March 29, 2011, http://fincher.house.gov/press-release/fincher-noem-call-epa-abandon-unreasonable-dust- standards; a joint letter from 75 House Members, September 27, 2010, http://agriculture.house.gov/letter/letter-epa-national-ambient-air-quality-standards-naaqs-particulate-matter-dust; a joint letter from 21 Senators, July 23, 2010, http://grassley.senate.gov/about/upload/Agriculture-07-23-10-dust-letter-to-EPA-signed-version-doc.pdf; an August 5, 2010, joint letter from former Senators Kent Conrad and Byron Dorgan and former Representative Earl Pomeroy. See also CRS Report R41622, *Environmental Regulation and Agriculture*, coordinated by Megan Stubbs.

[102] The Department of the Interior, Environment, and Related Agencies Appropriations Act, 2012 (H.R. 2584, Title IV Section 454) as reported by the House Committee on Appropriations on July 19, 2011. From July 25, 2011, to July 28, 2011, the House considered H.R. 2584 as reported July 19, 2011, but the House floor debate was suspended.

[103] See CRS Report R42332, *Environmental Protection Agency (EPA) FY2012 Appropriations*, by Robert Esworthy, and CRS Report R41979, *Environmental Protection Agency (EPA) FY2012 Appropriations: Overview of Provisions in H.R. 2584 as Reported*, by Robert Esworthy.

[104] For example, see U.S. Senate Committee on Environment and Public Works, Subcommittee on Clean Air, Climate Change, and Nuclear Safety, *Implementation of the Existing Particulate Matter and Ozone Air Quality Standards*, November 10, 2005.

[105] *Whitman v. American Trucking Associations*, 531 U.S. 457 (2001). Along with deciding issues specific to PM and ozone, the Court ruled unanimously that costs could not be considered in setting primary (health based) NAAQS.

INDEX

A

D

E

F

G

H

I

J

K

L

M

N

T

U

V

W

Y